Instructor Resources

Glencoe Carpentry
& BUILDING CONSTRUCTION

New York, New York   Columbus, Ohio   Chicago, Illinois   Peoria, Illinois   Woodland Hills, California

Reviewers:

Michael H. Frederick
Vocational Building Trades Instructor
Walker Career Center
Indianapolis, Indiana

Kathy C. Swan, Ph.D.
United Brotherhood of Carpenters
 and Joiners of America
Tacoma, Washington

Internet Disclaimer

The Internet listings throughout this guide provide a source for extended information related to the text. We have made every effort to recommend sites that are informative and accurate. However, these sites are not under the control of Glencoe/McGraw-Hill, and, therefore, Glencoe/McGraw-Hill makes no representation concerning the content of these sites. We strongly encourage instructors to preview Internet sites before students use them. Many sites may eventually contain links to other sites that could lead to exposure to inappropriate material. Internet sites are sometimes "under construction" and may not always be available. Sites may also move or have been discontinued completely by the time you or your students attempt to access them.

The McGraw-Hill Companies

Copyright © 2004 by Glencoe/McGraw-Hill, a division of The McGraw-Hill Companies. All rights reserved. Permission is granted to reproduce the material contained herein on the condition that such material be reproduced only for classroom use; be provided to students, instructors, or families without charge; and be used solely in conjunction with *Carpentry & Building Construction*. Any other reproduction, for use or sale, is prohibited without prior written permission of the publisher, Glencoe/McGraw-Hill.

Send all inquiries to:
Glencoe/McGraw-Hill
3008 W. Willow Knolls Drive
Peoria, IL 61614-1083

ISBN 0-07-822704-6 (Instructor Resource Guide)

ISBN 0-07-822702-X (Student Edition)
ISBN 0-07-825347-0 (Instructor Productivity CD-ROM)
ISBN 0-07-822703-8 (Carpentry Applications)
ISBN 0-07-825352-7 (Safety Guidebook)
ISBN 0-07-825353-5 (Carpentry Math)

Printed in the United States of America.

1 2 3 4 5 6 7 8 9 10   009   06 05 04 03 02

Instructor Resource Guide

Table of Contents

Program Overview 4
  Highlights of the six components of the *Carpentry & Building Construction* program.

Managing a Carpentry and Building Construction Program 11
  Resources and helpful information on topics such as technical advisory committees, motivation, meeting special needs of students, and assessment.

Safety First 29
  Safety handouts and a safety test.

Student and Professional Organizations 39
  Handouts describing student and professional organizations and the opportunities they provide.

Workplace Skills 49
  Handouts on topics such as work-based learning, finding a job, successful employee characteristics, teamwork, all aspects of industry, and entrepreneurship.

Computer and Internet Use 87
  Handouts to help students use the computer and the Internet effectively.

Instructional Plans 95
  Instructional plans keyed to each of the 45 chapters of the student edition.

Chapter Tests and Answer Key 189
  Tests keyed to each of the chapters in the student edition, plus an answer key.

Answer Key for Carpentry Math 328
  Answers to activity sheets provided in *Carpentry Math*, a supplement to the student edition.

Student Edition

ISBN 0-07-822702-X

Provides a comprehensive approach to residential and light commercial construction. Forty-five chapters cover the entire construction process from planning, estimating, and scheduling through the finishing touches.

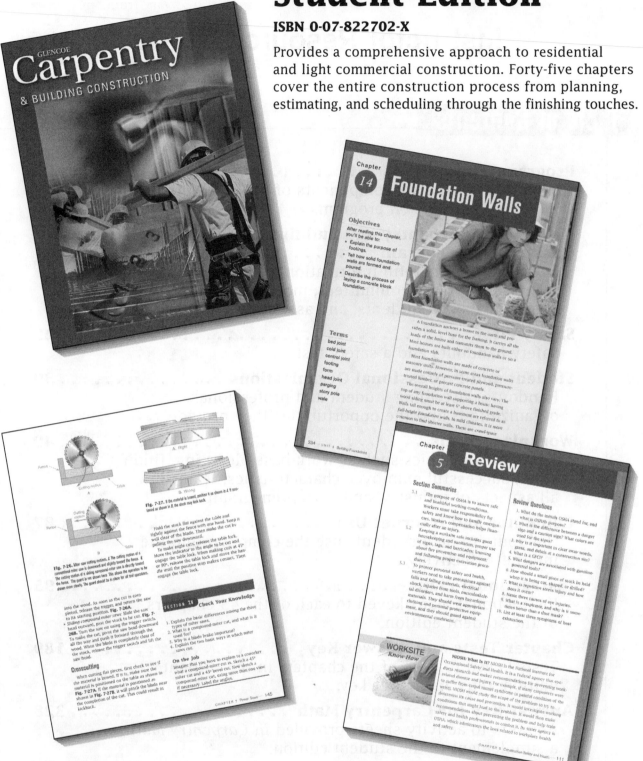

Check Your Knowledge questions reinforce key concepts in each section.

On the Job activities apply key concepts to practical situations.

Chapter Review provides section summaries and review questions.

Worksite Know-How presents information on job skills, tools, equipment, trade organizations, and health and safety issues.

Carpentry & Building Construction Instructor Resource Guide
Copyright © Glencoe/McGraw-Hill

Up-to-date drawings enhance the text with concrete details. For example, framing illustrations show precisely how members are joined.

From Another Angle highlights carpentry work practices that vary from one part of the country to the other.

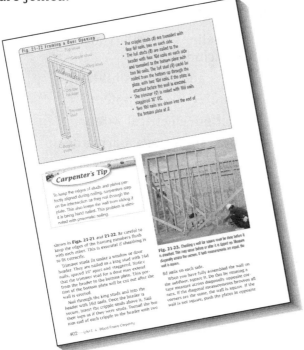

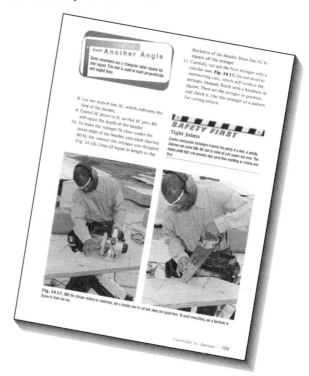

Carpenter's Tip offers workplace tips that save time and labor without sacrificing accuracy or quality.

Safety First emphasizes safety on the job, stressing work practices that require special attention to ensure personal safety on the worksite.

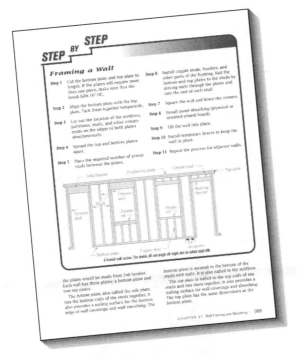

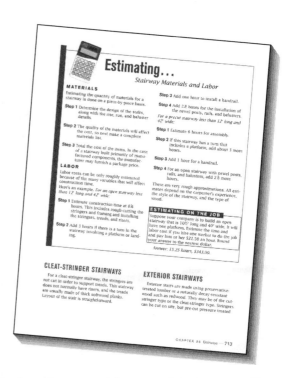

Step-by-Step presents the clear sequence of techniques needed to complete a given carpentry task.

Estimating provides step-by-step practice in using formulas to prepare material and labor estimates.

Safety Guidebook

ISBN 0-07-825352-7

Provides guidelines for safety and health on the job, including

- Rules, rights and responsibilities
- Personal protective equipment
- Fire protection, electrical safety
- Color codes, signs, tags, barricades
- Fall prevention
- Use of tools and equipment
- Handling and storage of materials

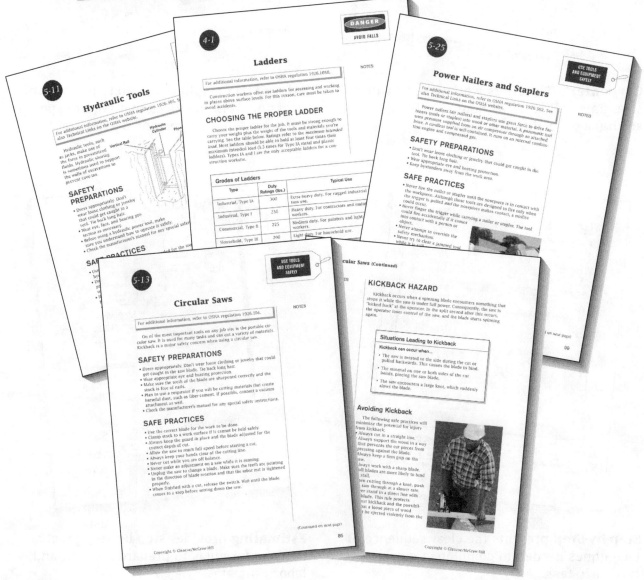

Carpentry & Building Construction Instructor Resource Guide
Copyright © Glencoe/McGraw-Hill

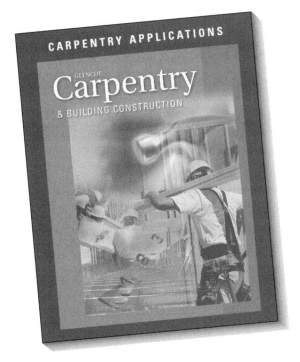

Carpentry Applications

ISBN 0-07-822703-8

Provides step-by-step instructions and illustrations for construction skills a carpenter needs. Application sheets address the following topics

- Wall framing, roof framing, and floor framing
- Building foundations
- Windows, doors, and siding
- Wiring, insulation, drywall, painting, flooring, tile, fireplaces, and decks
- Stairs, molding and trim, cabinets and countertops, and wall paneling
- Using tools and equipment
- Building permits, site plans, and building plans
- Safety awareness
- Design and problem solving
- Careers

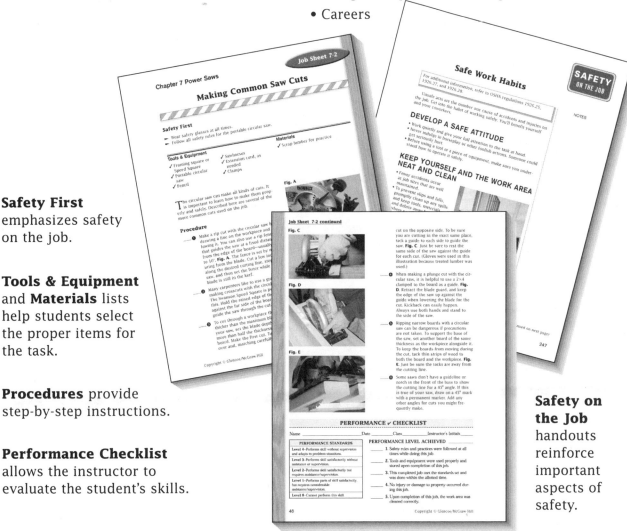

Safety First emphasizes safety on the job.

Tools & Equipment and **Materials** lists help students select the proper items for the task.

Procedures provide step-by-step instructions.

Performance Checklist allows the instructor to evaluate the student's skills.

Safety on the Job handouts reinforce important aspects of safety.

Carpentry & Building Construction Instructor Resource Guide
Copyright © Glencoe/McGraw-Hill

Carpentry Math

ISBN 0-07-825353-5

Provides exercises keyed to the student text chapters for

- Calculating fractions, decimals, and percentages
- Converting rules, scales, and rods
- Writing and using ratios
- Estimating materials

Practice Exercises provide detailed examples for solving problems.

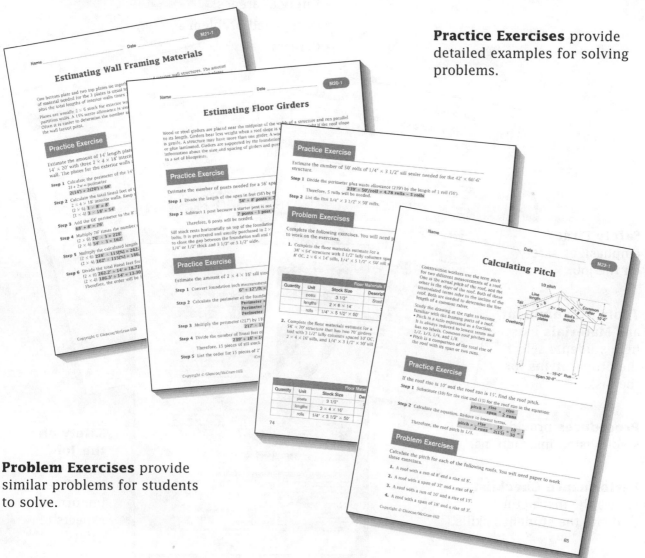

Problem Exercises provide similar problems for students to solve.

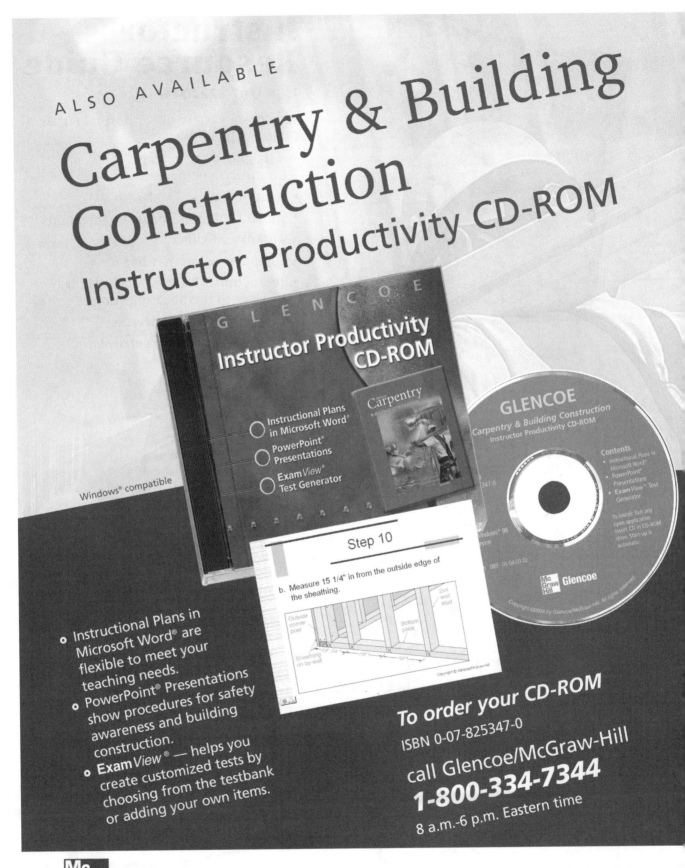

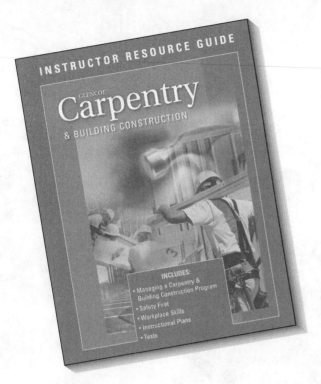

Instructor Resource Guide

ISBN 0-07-822704-6

Provides a variety of support materials. Contents include

- Managing a *Carpentry & Building Construction* Program
- Safety First
- Student & Professional Organizations
- Workplace Skills
- Computer & Internet Use
- Instructional Plans, which include answers to textbook questions
- Chapter Tests & Answer Key
- Answer Key for *Carpentry Math*

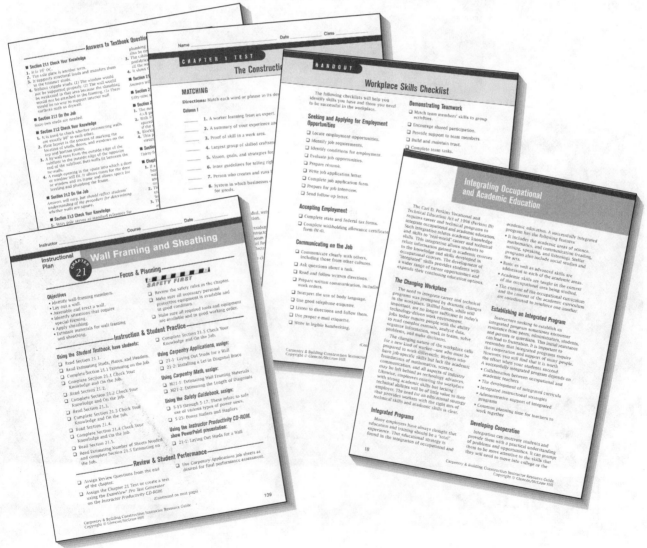

Carpentry & Building Construction Instructor Resource Guide
Copyright © Glencoe/McGraw-Hill

Managing a Carpentry and Building Construction Program

Contents

Motivating Students	12
Resources	13
Perkins III Legislation	15
Technical Advisory Committee	17
Integrating Occupational and Academic Education	18
Special Populations	21
Meeting Special Needs	23
Assessment	25
Assessment Strategies	28

Motivating Students

Motivation is an essential element in learning. It is a state of mind that creates the desire to learn, to solve problems, to create, and to succeed. Learners who are motivated and interested in their work are also less likely to present behavior problems.

Keys to Increasing Motivation

Four factors are often considered the keys to increasing motivation: attention, relevance, confidence, and satisfaction.

Increasing Attention

Strategies that create a sense of curiosity or adventure are good ways to get students' attention. Demonstrations and short discussions can spark and maintain students' interest.

Problem-solving activities often activate a sense of curiosity because students must seek knowledge or tap into their own knowledge in order to solve the problem. Good programs provide a wide variety of problem-solving activities from which to choose.

Increasing Relevance

Students will naturally be motivated to learn things that are relevant to their needs, interests, or goals. Required learning is more challenging. It helps to relate the subject matter to the real world with examples from the community or with products that are familiar to students.

Relevance relates to many other factors as well, including the way something is learned. For example, some students do best if allowed to participate in goal setting and to have some control over how the goals are achieved. Other students thrive by participating in groups that cooperate toward the achievement of a goal.

Increasing Confidence

Confidence relates to learners' beliefs about themselves and their ability to learn, which may be independent of what they can really do or not do. One way to increase a student's chance of success is to assign activities with an increasing level of difficulty, beginning with the most appropriate level possible for each student. Success leads to more success. Another way is to point out success due to effort rather than luck or an easy assignment.

Increasing Satisfaction

This category relates to how learners feel about their accomplishments. One way to improve this is to allow students to become increasingly independent in learning and practicing a skill. Have students learn new skills under low-risk conditions but practice well-learned tasks under realistic conditions. Demonstrations are a good way to introduce new skills in a low-risk way.

Intrinsic vs. Extrinsic Motivation

Motivators are the things that inspire learners to learn. They usually fall into two categories: intrinsic and extrinsic.

Intrinsic motivators lead to motivation from within each individual. Intrinsically motivated learning brings a sense of satisfaction and accomplishment. Intrinsic motivators vary according to each person's preferences and desires.

With *extrinsic* motivators, learning is motivated by external forces. The rewards result from the achievement but aren't directly related to it. For example, rewarding students with a party after they have completed a certain number of activities could be considered an external reward.

Consider extrinsic rewards carefully. Some may be perceived as controlling, which may stifle a learner's natural interest in the learning task. Rewards can mean a variety of different things to each learner, and each meaning can have different motivational and learning consequences.

Resources

Organizations

American Society of Safety Engineers
http://www.asse.org
Represents professionals who work in safety-related fields.

Association for Career and Technical Education (ACTE)
http://www.avaonline.org
A national education association that helps prepare youths and adults for careers.

American Subcontractor Association, Inc. (ASA)
http://www.asaonline.com
Represents subcontractors and offers information on government policies and resources that relate to construction issues.

APA-Engineered Wood Association
http://www.apawood.org
Researches and promotes engineered wood products. They also inspect members' engineered wood and certify that it meets their standards.

Associate General Contractors of America (AGC)
http://www.agc.org
Represents members of the construction industry and seeks to represent construction industry interests.

Center to Protect Workers' Rights (CPWR)
http://www.cpwr.com
Researches and informs workers about health and safety issues, as well as related economic issues.

International Conference of Building Officials (ICBO)
http://www.icbo.org
Offers the latest building code information.

Light Gauge Steel Engineers Association (LGSEA)
http://www.lgsea.com
A network of engineers who seek to make light gauge steel a preferred material for construction.

McGraw-Hill Construction Information Group (CIG)
http://construction.com
Provides information to construction businesses and professionals through the brands F.W. Dodge, Sweet's, Engineering News-Record, and Architectural Record.

National Association of Homebuilders (NAHB)
http://www.nahb.com
A federation of builders associations that works with the government, public, and business to provide quality homebuilding to consumers.

National Association of Women in Construction (NAWIC)
http://www.nawic.org
Works to further women's success in the construction industry.

National Roofing Contractors Association (NRCA)
http://www.nrca.net
Represents roofing contractors and offers information for businesses, government, and the public.

SkillsUSA-VICA
http://www.skills.usa.org
A national organization that helps train students for technical, skilled, or service occupations through activities and contests.

U.S. Department of Labor
http://www.dol.gov
Provides information and regulations about labor.

U.S. National Labor Relations Board (NLRB)
http://www.nlrb.gov
Ensures that U.S. labor laws are followed by employees and employers. It also provides information about labor laws.

United Brotherhood of Carpenters of America (UBC)
http://www.carpenters.org
A union that represents carpenters and other construction workers.

RESOURCES *(Continued)*

Material & Equipment Suppliers

DBI Sala Safety Equipment
http://www.salagroup.com
Produces rescue equipment and personal fall protection, such as safety harnesses.

Delta Machinery
http://www.deltawoodworking.com
Manufactures quality woodworking tools.

Grainger Industrial Supply
http://www.grainger.com
Sells a variety of products to business and industry.

Hilti
http://www.hilti.com
An international supplier of construction-related products.

Lowe's Home Improvement Warehouse
http://www.lowes.com
Discount retailer of home improvement supplies.

The Home Depot
http://www.homedepot.com
Discount retailer of home improvement supplies.

Porter Cable
http://www.porter-cable.com
Produces quality professional power tools.

Safetyhouse Signs, Inc.
http://www.safehousesigns.com
Provides a variety of signs that alert people to possible hazards in the workplace.

Simpson Strong-Tie Connectors
http://www.strongtie.com
Produces a variety of connectors for residential and commercial construction use.

USG Corporation
http://www.usg.com
Sells construction products, primarily products which contain gypsum.

Publications

Daily Journal of Commerce
http://www.djc-or.com
Features articles about design and construction.

Fine Home Building
http://www.taunton.com/finehomebuilding/index.asp
Features information on building techniques, remodeling, and other aspects of home building.

Frame Building News
http://www.framebuildingnews.com
A publication of the National Frame Builders Association that offers the latest framing-related news.

Journal of Light Construction
http://www.JLConline.com
Offers information on light construction and suggestions about running a light construction business.

Tech Directions
http://www.techdirections.com
Provides information for trade and technology education professionals.

TIES Magazine
http://www.tiesmagazine.org
Supports technology education and the integration of math, science, and technology.

The Internet listings provided here are a source for extended information related to our text. We have made every effort to recommend sites that are informative and accurate. However, these sites are not under the control of Glencoe/McGraw-Hill and, therefore, Glencoe/McGraw-Hill makes no representation concerning the content of these sites. We strongly encourage instructors to preview Internet sites before students use them. Many sites may eventually contain links to other sites that could lead to exposure to inappropriate material. Internet sites are sometimes "under construction" and may not always be available. Sites may also move or have been discontinued completely by the time you or your students attempt to access them.

Perkins III Legislation

An important source of funding is the Carl D. Perkins Vocational and Technical Education Act of 1998 (Perkins III). It provides support for career and technical education programs that prepare students for further learning and for a wide range of high-skill, high-wage careers. Perkins III focuses on high-quality programs that

- **Integrate occupational and academic education.** Solving the problems facing industry involves the use of applied communications, mathematics, and science. See pages 18-20 for more on this topic.

- **Provide students with strong experience in, and understanding of, all aspects of industry.** When occupational skills are limited to narrowly defined job tasks, it is difficult to integrate advanced academic skills. Covering all aspects of the industry provides a more complete academic context. See pages 66-74 for information and activities for applying the aspects of industry.

- **Promote student attainment of challenging occupational and academic standards.** Perkins III helps ensure that students learn the skills and knowledge they need to meet academic state standards, as well as industry-recognized skill standards. Meeting both kinds of standards will increase student success on the job.

- **Address the needs of individuals who are members of special populations.** Students in special populations need equal access to activities. The goal is to address the needs of these individuals to prepare them for further learning and for high-skill, high-wage careers. See pages 21-24.

- **Involve parents and employers.** Through advisory committees, parents, students, teachers, representatives of business and industry, labor organizations, special populations, and other interested individuals are involved in developing, implementing, and evaluating career and technical education programs. See page 17 for information on technical advisory committees.

- **Provide strong linkages between secondary and postsecondary education.** Cooperation between secondary and postsecondary education institutions can be a win-win situation for all concerned. Articulation programs provide this link and support students in acquiring the skills they need for future success. Articulation agreements for transferring credit are usually between high schools or career-technical centers and community or technical colleges.

- **Develop, improve, and expand the use of technology.** Today's workforce is driven by advanced technology and fast-paced innovation. Experience with high technology and communications will prepare students to adapt to a continually changing workforce.

- **Provide professional development for teachers, counselors, and administrators.** Opportunities for professional development, such as workplace internships for business experience or training in all aspects of industry, strengthen career and technical education programs. Opportunities are provided for both occupational and academic personnel.

Tech Prep

Perkins III reauthorized Tech Prep, a program that prepares students for careers by combining two years of secondary education with two years of postsecondary education or an apprenticeship program (referred to as 2+2 articulation). Perkins encourages the use of work-based learning and new technologies in tech-prep programs. It also supports partnerships among business, labor, and higher education.

PERKINS III LEGISLATION (Continued)

Workforce Investment Partnership

Perkins III works in conjunction with the Workforce Investment Partnership Act of 1998. This legislation created a "one-stop" coordinated system for federal aid programs for vocational education and other types of education and job training. Programs under Perkins are partners in this one-stop delivery system. The most important aspect of the act is its focus on meeting the needs of business for skilled workers, as well as the training, education, and employment needs of individuals.

Accountability

To promote continuous program improvement, Perkins III created a performance accountability system. The system uses "core indicators" for annual levels of performance. Core indicators include

- Student attainment of academic and occupational skill proficiencies
- Student attainment of a secondary school diploma or its recognized equivalent, a proficiency credential in conjunction with a secondary school diploma, or a postsecondary degree or credential
- Placement in, retention, and completion of postsecondary education or advanced training, placement in military service, or placement or retention in employment
- Student participation in, and completion of, career and technical education programs that lead to nontraditional training and employment. *Nontraditional employment* refers to occupations or fields of work that employ less than 25 percent of people from one gender.

Funding Uses

Funds made available to an eligible recipient under this title may be used

- To involve parents, businesses, and labor organizations in the design, implementation, and evaluation of career and technical education programs
- To provide career guidance and academic counseling for students participating in career and technical education programs
- To provide work-related experience, such as internships, cooperative education, school-based enterprises entrepreneurship, and job shadowing that are related to career and technical education programs
- To provide programs for special populations
- For local education and business partnerships
- To assist career and technical student organizations
- For mentoring and support services
- For leasing, purchasing, upgrading or adapting equipment, including instructional aides
- For teacher preparation programs that assist individuals who are interested in becoming career and technical education instructors, including individuals with experience in business and industry
- For improving or developing new career and technical education courses
- To provide support for family and consumer sciences programs
- To provide career and technical education programs for adults and school dropouts to complete their secondary school education
- To provide assistance to students to find an appropriate job and continue their education
- To support nontraditional training and employment activities

Carpentry & Building Construction Instructor Resource Guide
Copyright © Glencoe/McGraw-Hill

Technical Advisory Committee

The most important resource a technical instructor has is an active local advisory committee. Such a committee is recommended or even mandated by many state career and technical education boards. The purpose of the advisory committee is to assist in the overall planning and implementation of a curriculum that meets the needs of the industry and leads to job placement of students.

Career and technical education programs need to collaborate with the business and labor community. The advisory committee provides this necessary link and can help

- Determine the occupational needs in your industry
- Set up long-term and short-term goals for the program
- Keep the curriculum current in terms of job requirements
- Give advice concerning work and health standards
- Create good rapport between businesses and the school's career and technical education program
- Find suitable work experiences
- Assist in securing job placements for students
- Assist in the evaluation of the program
- Provide in-service training for instructors

To organize an advisory committee, consider the following steps

1. Check with your school board for policies regarding advisory committees.
2. Secure the names of people who are involved in various aspects of carpentry and building construction, such as local builders and architectural firms, union apprenticeship training center coordinators, and graduates of your program currently working in the industry.
3. Write each potential advisory committee member a letter asking for his/her service.
4. When the replies are received, send another letter thanking them for their willingness to serve and give a time and location for the first meeting.

It is usually wise to ask advisory committee members to serve for a minimum of one year. That way, those members whose attendance is lacking or are noncontributing, may be dropped after one year. Remember to send a letter thanking them for their service and then secure someone else to take their place. Those members who have been active may be continued by mutual agreement. Committee membership should be rotated so that you are never working with a totally "new" committee.

Once the advisory committee has been formed, the first meeting is of vital importance. The meeting should be well organized and short in duration. Within three days following the meeting, a copy of the minutes should be sent to each advisory committee member.

The use of an advisory committee can be one of the most effective means of keeping the community informed, keeping courses up-to-date, and selling the school board on the importance of career and technical education.

Integrating Occupational and Academic Education

The Carl D. Perkins Vocational and Technical Education Act of 1998 (Perkins III) requires career and technical programs to integrate occupational and academic education. Such integration relates academic knowledge and skills to "real-world" career and technical skills. This integration allows students to relate information gained in academic courses to the knowledge and skills developed in occupational courses. The development of "integrated" skills provides students with a wider range of career opportunities and expands their continuing education options.

The Changing Workplace

The need to integrate career and technical programs was prompted by dramatic changes in the workplace. Skillful hands, while still important, are no longer sufficient in today's technology-driven work environment. Jobs today require people with the ability to read complex manuals, analyze data, organize information, work in teams, solve problems, and make decisions.

The changing nature of the workplace calls for a new kind of worker—one who must be prepared to work differently. Workers who have job-specific skills but lack the academic foundations of mathematics, science, communication, and all aspects of industry may be left behind as technology advances. Likewise, employees entering the workplace with strong academic skills but lacking technical abilities will be of little value to their employer. The need for an educational strategy that provides workers with the right mix of technical skills and academic skills is clear.

Integrated Programs

Many employers have always thought that education and training should be a "total" experience. That educational strategy is found in the integration of occupational and academic education. A successfully integrated program has the following features

- It includes the academic areas of science, mathematics, and communication (reading, writing, speaking, and listening). Stellar programs also include social studies and the arts.
- Basic as well as advanced skills are addressed in each of the academic areas.
- Academic skills are taught in the context of the occupational area being studied.
- The content of the occupational curriculum and the content of the academic curriculum are coordinated to reinforce one another.

Establishing an Integrated Program

Instructors seeking to establish an integrated program sometimes encounter resistance from peers, administrators, students, and parents or guardians. This initial resistance can lead to frustration. It is important to remember that integrated programs require the cooperation and support of many people. However, you will find that it is worth the effort when your students succeed. A successfully integrated program depends on

- Collaboration between occupational and academic teachers
- The development of integrated curricula
- Integrated instructional strategies
- Administrative support of integrated programs
- Common planning time for teachers to work together

Developing Cooperation

Integration can motivate students and provide them with a practical understanding of problems and opportunities. It can prompt them to be more attentive to the skills that they will need to move into college or the

INTEGRATING OCCUPATIONAL AND ACADEMIC EDUCATION *(Continued)*

workplace. Also, by fostering communication between academic and occupational instructors, it will increase cooperation.

The key element in an integrated program is the working relationship between academic and occupational instructors. To take the first step, you will need to make sure the academic instructors are fully aware of the goals and advantages of an integrated program. Here are some useful strategies

- Invite your fellow academic instructors to your classroom.
- Provide them with copies of your textbooks, instructional materials, courses of study or course outlines, and any vocabulary lists or special projects that you have students complete each term.
- Invite them to attend your advisory committee meeting.
- Invite them to participate in student organization activities.
- Visit your fellow academic instructors' classrooms upon invitation.
- Ask to see the academic textbooks your students are using.
- Work with the academic instructors as equal partners when planning curriculum.

These are only first steps in bridging the gap between occupational and academic education. You should continue the process by

- Engaging academic instructors in serious discussions regarding how each can be supportive of what the students are expected to learn
- Discussing how content can be applied in a relevant and meaningful way
- Familiarizing your academic colleagues with the related mathematics applications found in the *Carpentry & Building Construction* program. For example, see the "Estimating" features in the student edition and the worksheets in *Carpentry Math*, a supplement to the student edition.
- Encouraging your peers to incorporate academic applications into their teaching

- Gaining administrative support for the integration of career and academic curricula by involving administrators in the planning process

Persistence Pays Off!

In obtaining the cooperation of your fellow instructors, persistence is important. Your colleagues need to be convinced that an integrated program will be of benefit to everyone. They need to know that the aim of integration is to improve the communication, mathematics, and science skills of your students.

Remember, the key element of integration is bringing occupational and academic instructors together. Cooperation is needed for curriculum development, scheduling, team teaching practices, and the development of assessments. The success of an integrated program also depends on the cooperation and organization, as well as the support, of the administration. The following strategies will be of assistance:

- **Instructional material must be a natural part of course content.** Such material should not appear to be special or unusual. The materials should relate naturally to the subject matter and be "integrated" as a natural part of the course content.
- **Encourage teamwork.** Most student activities should be conducted in teams. This is important for hands-on activities as well as for program presentations. Teams help students develop interpersonal skills. They also help mirror the modern workplace, where problem solving in a team framework has been proven to be effective. Remember, though, that the use of teams may also create conflict and disagreements within the team. These are a normal part of the process of team building, and you should be prepared to help resolve them.
- **Encourage student entries in skills competitions.** This will help develop support for integration. It will also validate

INTEGRATING OCCUPATIONAL AND ACADEMIC EDUCATION *(Continued)*

your approach and help you develop relationships with academic instructors.

- **Encourage students to develop portfolios.** Instruct students to use their portfolios to organize information and photos related to types of training they have received, skills they have learned, transcripts of their grades, and student-produced structures.
- **Organize public presentations.** This is an effective public relations strategy. Such presentations can be made before professional organizations, conferences, and individual schools. Organize selected students into a "traveling team." These students describe the project work they completed. They can also present their portfolios.
- **Encourage simulation.** For example, you might ask a team of students to set up a proposal for building a structure. The team might develop a construction plan and make a formal presentation to school administrators, asking permission to carry out their plan to simulate carpentry and building construction at work.

Measuring Success

The integration of occupational and academic education usually results in an individual who is better prepared to enter the workforce. Students often make dramatic gains in achievement when they are given more challenging work that integrates occupational and academic education. The success of an integrated program can be measured by the following factors

- Expanding enrollment
- Fewer discipline problems
- Respect of your peers and administration
- Employers recruiting your students
- Student success in competitive events, such SkillsUSA Championships sponsored by SkillsUSA-VICA
- Requests by teachers, students, and community members to visit your program
- Identification of your program by the state as exemplary
- Requests to make presentations to other schools and at state, regional, or national conferences

Special Populations

The Carl D. Perkins Vocational and Technical Education Act of 1998 (Perkins III) was passed for the purpose of improving student achievement and preparing students for further learning and for a wide range of opportunities in high-skill, high-wage careers. This includes addressing the needs of members of special populations in order to make sure they have equal access to all activities provided in educational programs. The definition of *special populations* includes people with disabilities and other barriers to educational achievement, such as limited English proficiency (LEP) and learning disabilities.

Perkins III programs must submit plans for how they will identify and adopt strategies to overcome barriers that result in limited access to vocational-technical education programs. The plan also must include a description of how programs will be designed to enable members of special populations to meet required performance measures. School districts or technical colleges must track and report special population student attainment of performance measures.

Individuals with Disabilities

Perkins III uses the American Disabilities Act (ADA) definition of individuals with disabilities. Such an individual has a physical or mental impairment that substantially limits one or more major life activities, has a record of such an impairment, or is regarded as having such an impairment. Major life activities are those that an average person can perform with little or no difficulty such as walking, breathing, seeing, hearing, speaking, learning, and working. Educational services must be designed to meet individual needs, and students with disabilities must be educated with nondisabled students.

Individuals with Limited English Proficiency

An individual with "limited English proficiency" refers to a secondary school student, an adult, or an out-of-school youth, who has limited ability in speaking, reading, writing, or understanding the English language. In addition, one or both of the following apply.

- The individual's native language is a language other than English.
- The individual lives in a family or community environment in which a language other than English is the dominant language.

Learning Disabilities

In addition to the ADA, the Individuals with Disabilities Education Act (IDEA) provides for people with difficulties related to learning. The IDEA strengthens expectations and accountability for students with disabilities and bridges the gap that sometimes occurs between what students with disabilities learn and what is required in a regular classroom.

Under IDEA, a student with a disability may have a physical, emotional, learning, or behavioral problem that is educationally related and requires special education and related services. The categories include autism, deaf-blindness, developmental delay, emotional/behavioral disorder, hearing impairment, mental impairment, multiple impairment, orthopedic impairment, other health impairment, specific learning disability, speech and/or language impairment, traumatic brain injury, and visual impairment (including blindness).

Individualized Education Program (IEP)

The IDEA requires that each public school student who receives special education and related services for a disability must have an Individualized Education Program (IEP). The IEP creates an opportunity for instructors, parents, school administrators, related services personnel, and students (when appropriate) to work together to improve educational results for students with disabilities. The IEP guides the delivery of special education

SPECIAL POPULATIONS *(Continued)*

supports and services for the student with a disability.

The writing of each student's IEP takes place within the larger picture of the special education process under IDEA. Here are the steps taken to establish an IEP:

1. **Student is identified.** The state must identify, locate, and evaluate all students with disabilities in the state who need special education and related services. A school professional or parent may ask for an evaluation. Parents must approve the evaluation.
2. **Student is evaluated.** The evaluation must assess the student in all areas related to the student's suspected disability.
3. **Eligibility is decided.** A group of qualified professionals and the parents look at the student's evaluation results. Together, they decide if the student is a "child with a disability," as defined by IDEA. Parents may ask for a hearing to challenge the eligibility decision.
4. **Student is found eligible for services.** If the student is found to be a "child with a disability," as defined by IDEA, he or she is eligible for special education and related services. The IEP team meets to write an IEP for the student.
5. **IEP meeting is scheduled.** The school system schedules and conducts the IEP meeting.
6. **IEP meeting is held and the IEP is written.** The team gathers to talk about the student's needs and write the IEP. For the student to be successful, it is very important for the student's instructor to be involved in writing the IEP.
7. **Services are provided.** The school makes sure that the student's IEP is being carried out as it was written. Parents are given a copy of the IEP. Each of the student's teachers and service providers has access to the IEP and knows his or her specific responsibilities for carrying out the IEP. This includes the accommodations, modifications, and supports that must be provided to the student, in keeping with the IEP.
8. **Progress is measured and reported to parents.** The student's progress toward the annual goals is measured, as stated in the IEP.
9. **IEP is reviewed.** The student's IEP is reviewed by the IEP team at least once a year, or more often if the parents or school ask for a review.
10. **Student is reevaluated.** At least every three years the student must be reevaluated. The purpose of the evaluation is to find out if the student continues to be a "child with a disability," as defined by IDEA, and what the student's educational needs are. However, the student must be reevaluated more often if conditions warrant or if the student's parent or teacher asks for a new evaluation.

Students with Special Needs

An instructor's challenge is to ensure that all students learn as well as possible. In addition to choosing the most effective strategies for achieving learning objectives, planning for student diversity will enrich the learning experiences for all. Students vary widely in their level of experience with a subject area and in their abilities.

Students who are unresponsive to instruction may have serious learning disabilities that require attention or they may need only minor adjustments in order to succeed. See the chart titled "Meeting Special Needs" for helpful resources and strategies for learning and other types of disabilities.

Meeting Special Needs

SUBJECT	DESCRIPTION	TIPS FOR INSTRUCTION
Limited English Proficiency (LEP)	Some students speak English as a second language, or not at all. Customs and behavior of people in the majority culture may be confusing for some of these students. Cultural values may inhibit some students from full participation in the classroom.	• Keep in mind that students' ability to speak English does not reflect their academic ability. • Try to incorporate students' cultural experiences into your instruction. The help of a bilingual aide may be effective. • Include information about different cultures in your curriculum to help build students' self-image. • Avoid cultural stereotypes. • Encourage students to share their cultures in the classroom.
Behaviorally Disordered	Students with behavior disorders deviate from standards or expectations of behavior and impair the functioning of others and themselves. These students may also be gifted or learning disabled.	• Work for long-term improvement; do not expect immediate success. • Talk with students about their strengths and weaknesses, and clearly outline objectives and tell how you will help them obtain their goals. • Structure schedules, rules, room arrangement, and safety for a conducive learning environment. • Model appropriate behavior for students and reinforce proper behavior. • Adjust group requirements for individual needs.
Visually Impaired	Students who are visually disabled have partial or total loss of sight. Individuals with visual impairments are not significantly different from their sighted peers in ability range or personality. However, blindness may affect cognitive, motor, and social development.	• Modify assignments as needed to help students become more independent. • Teach classmates how to serve as guides for the visually impaired; pair students so sighted peers can assist in cooperative learning work. • Tape lectures and reading assignments for the visually impaired. • For the benefit of the visually impaired, encourage students to use their sense of touch; provide tactile models whenever possible. • Verbally describe people and events as they occur in the classroom for the visually impaired. • Limit unnecessary noise in the classroom.

MEETING SPECIAL NEEDS (Continued)

SUBJECT	DESCRIPTION	TIPS FOR INSTRUCTION
Hearing Impaired	Students who are hearing impaired have partial or total loss of hearing. Individuals with hearing impairments are not significantly different from their peers in ability range or personality. However, the chronic condition of deafness may affect cognitive, motor, social, and speech development.	• Provide favorable seating arrangements so hearing-impaired students can see speakers and read their lips (or interpreters can assist); avoid visual distractions. • Write out all instructions on paper or on the board; overhead projectors enable you to maintain eye contact while writing. • Avoid standing with your back to the window or light source.
Physically Challenged	Students who are physically disabled fall into two categories—those with orthopedic impairments (use of one or more limbs severely restricted) and those with other health impairments.	• With the student, determine when you should offer aid. • Help other students and adults understand physically disabled students. • Learn about special devices or procedures and if any special safety precautions are needed. • Allow students to participate in all activities including field trips, special events, and projects.
Gifted	Although no formal definition exists, these students can be described as having above average ability, task commitment, and creativity. They rank in the top five percent of their classes. They usually finish work more quickly than other students, and are capable of divergent thinking.	• Emphasize concepts, theories, relationships, ideas, and generalizations. • Let students express themselves in a variety of ways including drawing, creative writing, or acting. • Make arrangements for students to work on independent projects. • Use public services and resources, such as agencies providing free and inexpensive materials, community services and programs, and people in the community with specific expertise. • Make arrangements for students to take selected subjects early.
Learning Disabled	All learning disabled students have a problem in one or more areas, such as academic learning, language, perceptions, social-emotional adjustment, memory, or ability to pay attention.	• To help students focus, provide short breaks and try placing them near the front of the room. • Break down tasks and provide step-by-step prompts. • Provide assistance and direction; clearly define rules, assignments, and duties. • Allow for pair interaction during class time; use peer helpers. • Practice skills frequently. • Distribute outlines of material presented in class. • Maintain student interest with games. • Allow extra time to complete tests and assignments.

Assessment

How will you know if students have learned? You will need to look for the appropriate and visible symptoms of learning—a change in the students' level of knowledge, attitude, and skill. You will identify the appropriate levels of learning by what the students do and say. This process of collecting and analyzing important (or relevant) information to determine the effectiveness of your teaching and the students' learning is called *assessment*.

Why Assess Student Learning?

The only way to be confident that your students have acquired the competencies in your program is to have them perform each task so that you can evaluate their performance for mastery. Four major uses of tests are

- Diagnosing entry-level competency of students
- Providing immediate feedback during the learning process
- Assessing mastery of each task/competency
- Evaluating effectiveness of instruction

Diagnostic Role

Tests can diagnose strengths and weaknesses of students, particularly in the area of essential prerequisite skills and knowledge. Often, students already possess some of the knowledge and skills needed. On the other hand, some students will have very little entry-level competence. Tests can provide a tool for you to assess where an individual student is now so that an appropriate instructional plan can be developed.

Providing Feedback

Learning is greatly enhanced when the student is given immediate and continuous feedback about progress. Short, informal written or performance tests can be provided to determine if the major concept or task has been mastered correctly.

Assessing Mastery

If the competency is knowledge, a written test can be used. If it is a skill, a performance test may be required. Both written and performance tests are used for tasks involving a great many concepts or principles (knowledge).

Evaluating Instruction

If you have an unusually high number of students who fail to achieve mastery of certain competencies, the problem might be with the quality of instruction. Well-developed tests can help evaluate the effectiveness of materials and learning experiences being used in order to improve quality continually.

What Should Be Assessed?

There are basically three kinds of learning. These are often classified as

- **Knowledge.** Does the student possess the required knowledge (facts, concepts, principles, etc.) to perform a specific competency?
- **Skills.** Does the student possess the necessary coordination to perform the task/competency?
- **Attitude.** Will the student choose to perform the task on the job after learning to do it?

Distributed among the three kinds of learning are tasks/competencies in these primary skill areas

- Occupational skills
- Academic skills
- Employability skills

ASSESSMENT *(Continued)*

Assessing Knowledge

Written tests can effectively assess student knowledge. When building a written test, try applying these basic rules.

1. **Have a specific purpose for the test.** The purpose might be diagnosis, skill attainment, determining grades, etc.
2. **Use student performance objectives as your guide.** The properly written student performance objective will clearly state the kind of assessment items appropriate for sampling the desired behavior.
3. **Test what was taught.** There should be "performance agreement" among the performance component in the objective, the content taught in the lesson, and the performance being assessed in the test.
4. **Test critical points of learning.** Do not emphasize inconsequential details.
5. **Use a variety of written test items.** Some students do not handle true-false items very well. Other students have trouble with multiple-choice tests. Others do quite well with short-answer questions. Be fair to your students. Consider their strengths and weaknesses when designing your tests.

Assessing Skills

While assessment of knowledge can be handled by a written test, skill assessment requires performance testing. A performance test measures skills that require some type of action. A performance test determines whether or not the student can actually perform the task in a job-like setting to a minimum standard.

The following are some steps to take in developing and conducting performance tests.

1. **Determine exactly what should be tested.**
2. **Determine whether process, product, or both are critical.** Process is how the learner performs the task. The product is the end result.
3. **Design the assessment tasks.** The assessment task should be designed to require the student to demonstrate attainment of the performance objectives.
4. **Specify the performance criteria.** In most cases the criterion or standards components of the student performance objectives will be specific enough to serve as the criteria for the performance test.
5. **Determine how the performance criteria or test items will be scored.** Two common methods are used to evaluate the learner's degree of competence: a rating scale and a checklist. Performance checklists have several advantages over rating scales. They are easier to score and are also more objective than a rating scale.
6. **Determine the minimum acceptable score.** One approach that may make it easy is to include only essential process- and product-related items. If all of the items are essential, the learner must demonstrate achievement of each performance criterion.
7. **Draft the test.** Three sections are important to assembling a performance test: preparation, student instruction, and the scoring guide.
8. **Try out the test.** Try out the test with a colleague and one or two students before using it to evaluate all students.
9. **Conduct the performance test.** Prepare the work area and provide students with all necessary equipment, tools, materials, instruments, references, drawings, and other items that may be required to complete the test. The testing situation should require the same equipment, content, sequence, and conditions as used in the instruction.
10. **Plan for feedback.** Provide students with an opportunity to ask questions and discuss their experiences after testing is completed. Discuss with them their individual strengths and weaknesses (feedback and follow-up).

ASSESSMENT (Continued)

Assessing Attitudes

Students can be tested to find out if they have mastered the principles behind desired attitudes. However, whether or not they will follow through as desired is a personal choice that is not easy to assess. On the other hand, learners can be observed for visible signs of adopting attitudes or values; for example, the appreciation of excellent workmanship, effective human relations, and professional and ethical behavior.

Reporting Assessment Results

As a classroom teacher, you will have to meet the grading and reporting requirements set by your school district. As you do that, keep the following points in mind.

- For performance assessment, grading and reporting should be done in reference to learning criteria, never on the curve. If the criterion component in the student performance objective is properly written, there should be no question in your mind as to how well each individual student has performed and what grade should be recorded for that performance. In reality, if each student meets the performance standard indicated in the criterion component of the objective, every student in the class should receive an "A".

- A single letter grade does not provide sufficient information about a student's performance in academic skills, occupational skills, and employability skills. Oral quizzes, written reports, job sheets, performance checklists, and chapter or unit tests work together to provide an accurate picture of a student's progress.

Assessment Materials in the *Carpentry & Building Construction* Program

The *Carpentry & Building Construction* program provides many types of materials that can be used for assessment. Those materials are described here. To order any of these components, call 1-800-334-7344 for customer service.

Student Edition (ISBN 0-07-822702-X)

Each chapter of the student edition provides a variety of tools that can be used for assessment. Each section closes with Check Your Knowledge questions and an On the Job activity. Each chapter closes with ten review questions. Estimating features focus on mathematics and include an Estimating on the Job activity for students to complete.

Carpentry Applications (ISBN 0-07-822703-8)

Most applications conclude with a Performance Checklist. These can be used for formative assessment as students practice and acquire new skills. You may want to have students repeat the checklist for each application until they have reached a level that you consider to be acceptable, which can then be used as the final, summative assessment.

Carpentry Math (ISBN 0-07-825353-5)

Activity sheets provide a variety of Practice Exercises and Problem Exercises that relate to topics in the student edition. Students apply basic math skills such as calculating measurements, fractions, and prices. Completed exercises can be used for assessing students' mathematical skills and their abilities to apply these skills to on-the-job situations.

Instructor Resource Guide (ISBN 0-07-822704-6)

This guide provides a test for each chapter of the student edition and a safety test for general power tool safety. In addition, handouts in various sections of this guide ask students to write answers or provide other types of input.

Instructor Productivity CD-ROM (ISBN 0-07-825347-0)

An electronic test generator includes expanded versions of the chapter tests printed in the instructor resource guide.

Assessment Strategies

STRATEGIES	ADVANTAGES	DISADVANTAGES
Objective Measures • Multiple choice • Matching • Item sets • True/false	• Reliable, easy to validate • Objective, if designed effectively • Low cost, efficient • Automated administration • Lends to equating	• Measures cognitive knowledge effectively • Limited on other measures • Not a good measure of overall performance
Written Measures • Essays • Restricted response • Written simulations • Case analysis • Problem-solving exercises	• Face validity (real life) • In-depth assessment • Measures writing skills and higher level skills • Reasonable developmental costs and time	• Subjective scoring • Time consuming and expensive to score • Limited breadth • Difficult to equate • Moderate reliability
Oral Measures • Oral examinations • Interviews	• Measures communications and interpersonal skills • In-depth assessment with varied stimulus materials • Learner involvement	• Costly and time consuming • Limited reliability • Narrow sample of content • Scoring difficult, need multiple raters
Simulated Activities • Role play • Computer simulations	• Moderate reliability • Performance-based measure	• Costly and time consuming • Difficult to score, administer, and develop
Portfolio and Product Analysis • Work samples • Projects • Work diaries and logs • Achievement records	• Provides information not normally available • Learner involvement • Face validity (real life) • Easy to collect information	• Costly to administer • Labor and paper intensive • Difficult to validate or equate • Biased toward best samples or outstanding qualities
Performance Measures • Demonstrations • Presentations • Performances • Rubrics • Observation	• Job-related • Relatively easy to administer • In-depth assessment • Face validity	• Rater training required • Hard to equate • Subjective scoring • Time consuming if breadth is needed
Performance Records • References • Performance rating forms • Parental rating	• Efficient • Low cost • Easy to administer	• Low reliability • Subjective • Hard to equate • Rater judgment
Self-evaluation • Rubrics • Checklists	• Learner involvement and empowerment • Learner responsibility • Measures dimensions not available otherwise	• May be biased or unrealistic

Safety First

Contents

- Classroom/Lab Safety Checklist 30
- Student Emergency Information Sheet 31
- Student Safety Performance Record 32
- General Power Tool Safety 33
- General Power Tool Safety Test 35
- Answer Key for General Power Tool Safety Test 37
- Developing a Fire Emergency Plan 38

Classroom/Lab Safety Checklist

- ❏ A comprehensive safety program is in place and includes demonstrations, written tests, and performance assessments for safe operation.
- ❏ Documentation of parental safety authorization.
- ❏ Student safety records and accident report forms are maintained for each student.
- ❏ Equipment guards are secured and in good condition.
- ❏ Number of students is appropriate for activities.
- ❏ Appropriate fire extinguishers are in place.
- ❏ Safety zones are marked near equipment.
- ❏ Compressed gas cylinders are properly secured.
- ❏ Appropriate ventilation for activities.
- ❏ Proper exhaust/intake for activities.
- ❏ Proper storage of flammable and/or combustible liquids.
- ❏ Tools are in good condition and stored properly.
- ❏ Properly sanitized personal protective equipment is provided.
- ❏ Gas and electrical shutoff valves/switches are readily accessible.
- ❏ First aid kit is adequately stocked and readily accessible.
- ❏ Ladders are in good condition and provided where needed.
- ❏ Heavy items are stored in lower shelves.
- ❏ Storage area is neat and clean.
- ❏ Power machines are properly lubricated and in good condition.
- ❏ Automated equipment: Cables are neat, wires are not frayed, limit switches and emergency switches are operational and unobstructed.
- ❏ Hand tools are sharp, properly stored, operational, and free of dirt and grease.
- ❏ Oily rags are disposed of in self-closing, noncombustible containers.
- ❏ Lab rules and signs enforcing safe operation of machines are posted (i.e., jewelry removal).
- ❏ A noise assessment and reduction plan is in place.
- ❏ Other _____

Signature of evaluator: _____ Date: _____

HANDOUT

Student Emergency Information Sheet

Student's Name _____

List person(s) to contact in case of emergency.

Name: _____ Relationship to Student: _____

Home Address: _____

Home Phone: _____ Work Phone: _____

Wireless Phone: _____

Name: _____ Relationship to Student: _____

Home Address: _____

Home Phone: _____ Work Phone: _____

Wireless Phone: _____

List any special needs (e.g., medications, allergies, disabilities, etc.).

_____

_____

If you cannot be located in case of serious injury to this student, indicate a physician for emergency treatment:

_____

_____

Preferred Hospital: _____

Signature of Parent or Legal Guardian: _____

Date: _____

Carpentry & Building Construction Instructor Resource Guide
Copyright © Glencoe/McGraw-Hill

HANDOUT

Student Safety Performance Record

School: _____ Teacher: _____

Program: _____ Period: _____ Year: _____

_____ has observed safe operating procedures,
(Student's Name)
has passed required safety exams, and is permitted to operate the following items.

Tools or Equipment	Date Completed	
	Written Safety Assessment Passed	Performance Safety Assessment Passed

Student's Signature _____ Date: _____

Instructor's Signature _____ Date: _____

Carpentry & Building Construction Instructor Resource Guide
Copyright © Glencoe/McGraw-Hill

HANDOUT

General Power Tool Safety

The following are general safety rules for power tools. Check the manufacturer's manual for any special safety instructions.

NOTES

✔ Assure Cord Safety

- Avoid damaging cords or hoses. Never carry a tool by the cord or hose. Never yank the cord to disconnect it from a receptacle.
- Always use the recommended extension cord size when using portable power tools.
- Take steps to ensure that power cords won't create a tripping hazard.

✔ Use Tools Properly

- Avoid accidental starts. Do not hold your fingers on the on/off switch while carrying a plugged-in tool. Also, be sure the tool's switch is in the "off" position before plugging the tool in.
- Unplug or disable tools when not using them, before servicing or cleaning them, and when changing or adjusting accessories, such as blades, bits, and cutters.
- Always check lumber for knots, splits, nails, and other defects before machining it.
- Use guards on power equipment and be sure they are installed correctly. Standard safety guards must never be removed when using a portable power tool.
- Keep your fingers away from the cutting edges of tools. Don't try to hold small stock while it is being cut, shaped, or drilled. Secure it with clamps or in a vise whenever possible. This frees both hands to operate the tool.

✔ Work Safely

- Pay attention to the job. Always keep your eyes focused on where the cutting action is taking place. Don't talk with others while you work. Never talk to or interrupt anyone else who is using a power tool.
- Always use a brush, not your hand, to clean sawdust away from a power tool.
- Be sure to keep good footing and maintain good balance when operating power tools.
- Loose clothing or jewelry can become caught in moving parts. Wear proper apparel for the task. Tie back long hair.
- Keep work areas well lighted.

(Continued on next page)

General Power Tool Safety *(Continued)*

NOTES

- Never leave tools or materials on any piece of equipment while it is in use. This is especially important with table saws.
- When finished with a power tool, wait until the blade or cutter has come to a complete stop before walking away. Make all adjustments with the power off and the machine at a dead stop.

✔ Take Care of Tools

- Report strange noises or faulty operation of machines to your instructor or supervisor.
- Remove all damaged tools from use and tag them: "Do not use."
- Store electric tools in a dry place when not in use. Do not use them in damp or wet locations unless they are approved for those conditions.

Name _____ Date _____ Class _____

General Power Tool Safety Test

COMPLETION

Directions: In the blank provided, write the word or words that best complete the statement or answer the question.

1. When tools are damaged, they should be removed and tagged with the words:

   "Do not _____."

2. Do not _____ a power cord to disconnect it from a receptacle.

3. Strange noises and faulty machine operation should always be reported to your

   _____.

4. Never _____ or talk to someone operating a power tool.

5. Before machining lumber, always check for _____, knots, splits, and other defects.

6. Before walking away or making adjustments to a power tool, make sure the power is off

   and the machine is at a dead _____.

7. You should never carry a tool by the _____.

8. Instead of trying to hold small stock while it is being cut, shaped, or drilled, secure it

   with _____ or a vise.

9. Always use the recommended extension cord _____ when using portable power tools.

10. Do not use power tools under damp or _____ conditions unless they are approved for those conditions.

11. Always be sure the tool's switch is in the _____ position before plugging the tool in.

12. When changing or adjusting accessories, such as blades, bits, and cutters, always disable

    or _____ tools.

(Continued on next page)

Carpentry & Building Construction Instructor Resource Guide
Copyright © Glencoe/McGraw-Hill

GENERAL POWER TOOL SAFETY TEST (Continued)

13. Never leave materials or _____ on any piece of equipment—especially a table saw—while it is in use.

14. When using a portable power tool, never remove the standard safety _____.

15. Always take steps to ensure that power _____ won't create tripping hazards.

16. Always pay attention to the job. Don't _____ with others while you work.

17. To clearly see your work, the equipment, and other items, keep work areas well

_____.

18. When cleaning sawdust away from a power tool, always use a _____, not your hand.

19. When operating power tools, be sure to keep good footing and maintain good

_____.

20. Because loose clothing and jewelry can become caught in moving parts, wear proper

apparel. Tie back long _____.

Carpentry & Building Construction Instructor Resource Guide
Copyright © Glencoe/McGraw-Hill

Answer Key for General Power Tool Safety Test

1. use
2. yank
3. instructor (or supervisor)
4. interrupt
5. nails
6. stop
7. cord
8. clamps
9. size
10. wet
11. off
12. unplug
13. tools
14. guard(s)
15. cords
16. talk
17. lighted
18. brush
19. balance
20. hair

HANDOUT
Developing a Fire Emergency Plan

NOTES

Your lab should have a plan for use in a fire emergency. If your instructor has not explained it to you, ask about it. Be sure you know where all the exits are, where the fire extinguishers are kept, and what your responsibilities are in case a fire occurs. Then follow these steps to record the fire emergency plan.

1. On a separate piece of paper, draw a floor plan of your lab.
2. Locate all the fire exits and label them on the floor plan. If the lab has no direct exit to the outside, use a separate piece of paper and draw a map of your section of the school. Show the location of the lab, and draw arrows from the doors in the lab to the nearest fire exits.
3. Locate and label any windows that could also be used for escape. (Check to be sure the opening will be large enough for an adult to pass through.)
4. Locate all the fire extinguishers and label them on the floor plan. Indicate on the plan the class of fire for which each extinguisher can be used.
5. Locate places where flammable or explosive materials should be stored and label them. Be sure they are in a location far away from any source of heat and that their cabinet is fireproof.
6. Determine where you and your classmates should meet after you have left the building. This is important so that someone can check to be sure everyone has escaped safely.
7. Ask your instructor for procedures to follow if a fire occurs. If time allows, this may include such things as closing windows, turning off equipment, and grabbing the first aid kit. Someone should be responsible for reporting the fire. However, human safety is the most important consideration. Fires can spread quickly, and smoke can be just as deadly as flames. Your most important responsibility is to get out alive.
8. Be sure everyone has a chance to study the emergency plan. Then, with your instructor's approval, post it in a prominent place where it can be seen easily.
9. Be sure that you and your classmates are informed as to housekeeping duties that help prevent fires, proper handling of combustibles, how to treat burns, and where to report a fire.

Student and Professional Organizations

Contents

SkillsUSA-VICA 41

SkillsUSA Pledge, Creed, Colors 42

SkillsUSA Championships 43

Parliamentary Procedures 44

United Brotherhood of Carpenters and Joiners of America 45

Associated General Contractors of America 46

National Association of Home Builders 47

National Association of Women in Construction 48

Carpentry & Building Construction Instructor Resource Guide
Copyright © Glencoe/McGraw-Hill

SkillsUSA-VICA serves high school and college students and professional members who are enrolled in training programs in technical, skilled, and service occupations, including health occupations.

Purpose

The purpose of SkillsUSA-VICA is to prepare high-performance workers. Its motto is "preparing leadership in the world of work." SkillsUSA-VICA

- Provides experiences for students in leadership, teamwork, citizenship, and character development
- Builds and reinforces self-confidence, work attitudes, and communication skills
- Emphasizes total quality at work, high ethical standards, superior work skills, lifelong education, and pride in the dignity of work
- Promotes understanding of the free enterprise system and involvement in community service activities

History

SkillsUSA-VICA began in 1965 as the Vocational Industrial Clubs of America, Inc. (VICA). It was founded by students and teachers who were serious about their professions and saw the need for more training in the areas of leadership to complement their chosen vocation.

In 1967, VICA began holding competitive events (Skill Olympics). In 1981 VICA hosted the International Youth Skill Olympics, where VICA members joined 274 international contestants from 14 countries in 33 contests.

In 1994, the name Skill Olympics was changed to SkillsUSA Championships. In 1999, VICA officially changed its name to SkillsUSA-VICA.

Programs

SkillsUSA-VICA hosts local, state, and national competitions in which students demonstrate occupational and leadership skills. SkillsUSA-VICA programs also help to establish industry standards for job skill training in the classroom.

Two additional programs help prepare students for the world of work. The *Total Quality Curriculum* emphasizes the competencies and essential workplace basics skills. The *Professional Development Program* guides students through 84 employability skills lessons. These include goal setting, career planning, and community service.

SkillsUSA-VICA Chapters

Each chapter has an advisor, elects officers, and follows SkillsUSA-VICA regulations. Each chapter also develops an operating plan, which is called the program of work. This includes the activities in which members participate. SkillsUSA-VICA recommends an emphasis on activities such as leadership development, social development, community service, fund-raising, career development, skills competition, and public relations.

The national organization makes publications available to guide chapters in how to set up the program of work, as well as other issues concerning how to operate. For example, each chapter needs to know the types of memberships available, what officers to elect and what each officer's duties are, how to conduct the opening and closing ceremonies, clothing requirements, requirements for a club to conduct business legally, and the procedures to complete a business meeting. This includes how to use *parliamentary procedures*, which are the generally accepted rules for conducting business meetings.

Contact Information
SkillsUSA-VICA
www.skillsusa.org

SkillsUSA-VICA Pledge

Upon my honor, I pledge
- To prepare myself by diligent study and ardent practice to become a worker whose services will be recognized as honorable by my employer and fellow workers
- To base my expectations of reward upon the solid foundation of service
- To honor and respect my vocation in such a way as to bring repute to myself
- And further, to spare no effort in upholding the ideals of SkillsUSA-VICA

SkillsUSA-VICA Creed

- **I believe in the dignity of work.**

I hold that society has advanced to its present culture through the use of the worker's hands and mind. I will maintain a feeling of humbleness for the knowledge and skills that I receive from professionals, and I will conduct myself with dignity in the work I do.

- **I believe in the American way of life.**

I know our culture is the result of freedom of action and opportunities won by the founders of our American republic, and I will uphold their ideals.

- **I believe in education.**

I will endeavor to make the best use of knowledge, skills, and experience that I will learn in order that I may be a better worker in my chosen occupation and a better citizen in my community. To this end, I will continue my learning now and in the future.

- **I believe in fair play.**

I will, through honesty and fair play, respect the rights of others. I will always conduct myself in the manner of the best professionals in my occupation and treat those with whom I work as I would like to be treated.

- **I believe satisfaction is achieved by good work.**

I feel that compensation and personal satisfaction received for my work and services will be in proportion to my creative and productive ability.

- **I believe in high moral and spiritual standards.**

I will endeavor to conduct myself in such a manner as to set an example for others by living a wholesome life and by fulfilling my responsibilities as a citizen of my community.

SkillsUSA-VICA Colors

The colors red, white, blue, and gold represent the national SkillsUSA-VICA organization.

- Red and white represent the individual states and chapters.
- Blue represents the common union of the states and of the chapters.
- Gold represents the individual, the most important element of the organization.

HANDOUT

SKILLS USA CHAMPIONSHIPS

SkillsUSA-VICA provides local, state, and national competitions in which students demonstrate their occupational and leadership skills. These contests recognize the achievements of career and technical education students and encourage them to strive for excellence and pride in their chosen fields.

The national SkillsUSA Championships event is held annually. Thousands of students compete in over 70 occupational and leadership skill areas. Working against the clock and each other, the participants prove their expertise in skills such as technical drafting, electronics, and precision machining. There are also competitions in leadership skills, such as extemporaneous speaking and conducting meetings by parliamentary procedure.

The contests are planned by technical committees made up of representatives of labor and management. The national technical committee is assisted by local representatives of education and industry. Safety practices and procedures are judged and graded.

Students benefit from the competition no matter how they place in the finals. They learn more about their skills and often make future job contacts.

Carpentry

The contestants are required to complete a project specifically designed to test overall carpentry skills which include framing, exterior trim, and stair building. They are judged on the basis of accuracy, ability to read and interpret blueprints, workmanship, and the proper use of tools and equipment.

Job Interview

The contest is divided into three phases: completion of employment application, preliminary interview with receptionist, and an in-depth interview. The contestant's understanding of employment procedures he or she will face in applying for positions in the occupational areas in which he or she is training is evaluated.

TeamWorks

In the TeamWorks contest, students will work in a team of four to build a joint project, demonstrating preparation for employment and excellence and professionalism in the fields of residential carpentry, masonry, plumbing, electricity, and teamwork skills.

World Skills Competition

The World Skills Competition (WSC) is a biennial contest in skilled occupations. Through competition at the WSC, students learn what it means to be a competitor in the global market. The United States' top career and technical students test their skills against teams from other countries in an international showcase of occupational training. More than 600 young people from over 30 participating countries compete in over 40 skills and techniques competitions.

Contact Information
SkillsUSA-VICA
www.skillsusa.org

Carpentry & Building Construction Instructor Resource Guide
Copyright © Glencoe/McGraw-Hill

Parliamentary Procedures

Parliamentary procedures are the generally accepted rules used for business meetings. The purpose of the rules is to help a group make decisions in an orderly manner and to promote cooperation and harmony. *Robert's Rules of Order* is a commonly used guide to parliamentary procedures.

To make sure that issues are decided fairly, a quorum must be present. A *quorum* is usually defined as a majority of the members, unless otherwise defined by the organization's bylaws or other rules.

To keep order, only one issue can be considered at one time. No member may speak until recognized by the chair (group leader), and the chair should, of course, be strictly impartial.

Addressing Issues with a Motion

When a member has an issue to deal with, it must be presented in the form of a motion before it can be discussed. There are three steps for making a motion.
1. The member stands up, is recognized, and makes a motion. For example, "I move that we buy new uniforms, including hats."
2. Another member seconds the motion.
3. Without rewording it, the chair states the question being considered in the motion.

The members then debate the motion. After debate is complete, the chair puts the motion to a vote. After the vote is complete, the chair announces the results.

Voting on the Motion

Three basic methods are often used for voting.
- **Voice vote.** The chair says, "All in favor of the motion, say 'aye'" and "those opposed, say 'no.'"
- **Show of hands.** Members vote by raising their hand. For example, the chair might say, "All in favor, raise your right hand," and so on.
- **Rising vote.** Members stand to vote yes or no.

If a secret vote is necessary, a ballot can be used. A roll call vote can also be used. This places the vote of each member on record.

A majority vote usually decides if a motion is approved. *Majority* usually means more than one-half of the voting members. Some organizations require a two-thirds vote to pass a motion.

Amending the Motion

There are three ways to amend a motion. You can also amend the amendment.
1. You can move to amend by inserting words or paragraphs. For example, "I move to amend by inserting the phrase 'with the total cost no more than $500.'"
2. You can move to amend by striking (not deleting) words or paragraphs. For example, "I move to amend by striking out the words 'including hats.'"
3. You can move to amend by striking out and inserting words or paragraphs. You can even replace the entire motion. For example, "I move to amend by striking out 'buy new uniforms, including hats' and inserting the words 'rent a van for the national competition.'"

Four Types of Motions

- **Main.** A proposed action presented for consideration and decision.
- **Subsidiary.** These help resolve the main motion. Examples are an amendment and a referral.
- **Privileged.** Motions concerned with the rights of members. They take precedence over all other motions. Examples are to adjourn or to recess.
- **Incidental.** A type needed due to business being conducted at a meeting. One example is a point of order, which draws attention to a rule that is not being followed properly.

HANDOUT

United Brotherhood of Carpenters and Joiners of America

One group that strives to represent and support construction workers is the United Brotherhood of Carpenters and Joiners of America (UBC). The UBC is a union, a group of workers organized to improve working environments and benefits for their members.

Purpose

The UBC was founded by Peter J. Maguire in 1881 to unite and serve workers and their families. It continues to negotiate with companies and support legislation to benefit workers.

Membership

The UBC represents people who build houses, bridges, furniture, and commercial buildings. In addition to traditional carpenters, it includes drywall installers, mill workers, floor installers, lathers, tile workers, and other types of workers.

Members pay dues, which allows the organization to function. They can take part in elections and be hired by union companies. The union provides services that benefit these members.

Activities

A priority in the UBC is offering quality education. For example, the union provides a comprehensive apprenticeship program. The UBC and local construction companies work together to ensure that apprentices receive quality, hands-on training in carpentry. The cost of the training is paid for by the union and the participating company. Successful completion of the program results in "journeyman" status. Persons who possess trade skills but have not completed a union apprenticeship may take a test to determine whether they are qualified for union journeyman status. Journeymen earn full union wages and benefits.

The UBC is also involved with Job Corps, a program that allows economically disadvantaged young people the opportunity to participate in construction training. The Jobs Corps trains young people in trades such as brick masonry, carpentry, electrical wiring, and landscaping.

In addition, the UBC provides training to journey-level workers so that they can continually acquire the latest skills in their trade. Training is provided in training centers across North America.

Unions such as the UBC work to improve the quality of workers' lives, workplaces, and companies. The union negotiates contracts with the employees' companies and also aids workers in solving grievances. Companies benefit from the union by receiving well-trained workers who value their work.

Impact

The United Brotherhood of Carpenters and Joiners of America has helped shape many of the positive aspects of today's working environment. In its early history, the UBC worked with other unions to provide the eight-hour work day and better wages, benefits, and working conditions.

The UBC continues to work for these goals today. It is shaping its structure and efforts so that it can support the unique needs of today's workers.

Contact Information
The United Brotherhood of Carpenters
   and Joiners of America
www.carpenters.org

Associated General Contractors of America

The Associated General Contractors of America (AGC) is the largest construction trade association in the United States. AGC members are construction contractors and industry-related companies. AGC members can either be unionized or open shop. They seek to serve their members so that both public and private construction can be completed to high standards of quality.

The Three Tenets

Members of AGC are committed to three tenets. These tenets are skill, integrity, and responsibility. Through the development of these qualities, industry and its workers are able to improve construction techniques and provide even higher quality projects to the public and private sectors.

Education

One way in which the AGC develops these tenets is through education. AGC is supportive of joint labor-management apprenticeships. The association accredits high school and community college vocational programs to ensure quality trade education.

AGC also sponsors activities that showcase the appeal of careers in the construction industry. It sponsors student chapters and is supportive of contests such as SkillsUSA-VICA. Along with the Associated Schools of Construction (ASC), it sponsors the ASC-AGC National Student Competition, which encourages college students to excel in construction-related tasks. AGC provides a variety of scholarships which help qualified students learn the latest skills.

Legislative Work

AGC also works with legislators to ensure that the construction industry and its workers are effectively utilized by public and private projects.

Labor and Human Resources

AGC members can access a variety of information about labor and human resources through AGC. AGC provides information on wages, bargaining agreements, and labor laws. This service allows employers to respond fairly to their employees.

Contact Information
Associated General Contractors
   of America
www.agc.org

National Association of Home Builders

The National Association of Home Builders is an organization of companies that are involved in residential construction. They have joined together in order to share information about the latest construction technology, to advance government policies that promote home-building, and to offer consumers affordable, high-quality construction.

Education

NAHB provides information to its members, the public, and the government in various ways. They sponsor trade shows so that home builders can learn about the latest products. Members can then share this information with their customers, who can benefit from more affordable and useful building supplies. NAHB works to inform government representatives of the needs of home buyers and builders so that appropriate legislative action can be taken. Some of the following examples show the wide variety of ways that the NAHB seeks to improve the residential construction industry.

Student Chapters

NAHB sponsors student chapters that give high school and postsecondary students the opportunity to learn more about residential construction. Student members receive publications, opportunities to win awards, and discounts on trade shows and other NAHB sponsored events.

Some of the activities that chapters coordinate are construction site visits, campus building rehabilitation projects, tutoring groups, job fairs, and visits to local Home Building Association meetings.

Workforce Development

One branch of the NAHB is the Home Builder's Institute (HBI), an educational program that seeks to train workers for the construction industry.

One of its programs is Job Corps, which allows economically disadvantaged young people the opportunity to participate in construction training. The Jobs Corps trains young people in trades such as brick masonry, carpentry, electrical wiring, and landscaping. The program benefits both companies and workers because it provides young people with jobs and companies with dedicated, skilled entry-level workers.

NAHB Research Center

An independent subsidiary of the NAHB is the NAHB Research Center. This center seeks to use objective scientific data to further the quality of residential construction. The center is an information source for consumers, builders, and related industries.

Some NAHB Research Center projects include creating affordable housing, environmentally friendly housing, and housing which suits consumers with special needs.

Contact Information
National Association of Home Builders
www.nahb.org

The National Association of Women in Construction

The National Association of Women in Construction (NAWIC) serves women who work in the construction industry in a range of careers from the skilled trades to business ownership. The organization also supports educational programs for construction-related fields.

Purpose

The purpose of NAWIC is "to enhance the success of women in the construction industry." In order to fulfill this purpose, NAWIC promotes education, provides scholarships, develops partnerships, and presents awards that showcase women's contributions in the construction industry.

Educational Programs

The NAWIC Education Foundation offers a variety of opportunities for students to gain and improve construction-related skills. One program is the CAD/Design/Drafting Competition for high school students. Students create a solution to a "Design Problem" that includes at least a floor plan, exterior elevation, and a site plan showing the location of their design. The designs are judged for creativity, precision, and detail. The program encourages mentoring, the development of self-confidence, and an awareness of the construction industry.

Another activity sponsored by NAWIC is Women in Construction Week (WIC). Local chapters offer young women opportunities to participate in contests, to take tours of construction sites, and to attend Career Days on construction. Hands-on activities allow women to experience construction as a career choice. Young women learn that construction-related jobs can offer workers good wages, excellent benefits, opportunities for advancement, and a diverse work environment.

NAWIC Founders' Scholarship Foundation also helps women to gain more construction-related education. NAWIC offers scholarships for students already enrolled in undergraduate or construction-related degree programs.

Partnerships

NAWIC works with other construction and business groups to further women in construction. These partnerships focus on increasing opportunities for women's training, education, and business ownership. American Subcontractors Association (ASA), Associated Builders and Contractors (ABC), and Women's Business Enterprises Council (WBENC) are just a few of NAWIC's partners.

Women Build of Habitat for Humanity is another NAWIC partner. This program allows women to volunteer to build homes for other women in need of affordable housing. The program encourages volunteers with a variety of levels of experience, so women with construction experience can train other women who want to develop construction skills. The program gives women practical experience in construction and allows them to see the many roles that women do and can play in construction. These experiences work to enhance the success of women in construction.

Results

Through the efforts of determined women and groups such as NAWIC, women's involvement in construction has increased significantly. By developing contact with organizations such as NAWIC, you can learn about the rewards and benefits of careers in construction for women.

Contact Information
National Association of Women in Construction
www.nawic.org

Workplace Skills

Contents

Successful Employee Characteristics	50
Teamwork	52
Apprenticeship Training	55
Work-Based Learning	56
Workplace Skills Checklist	59
Working with Employers	61
Liability Issues	62
Work-Based Learning Agreement	63
Workplace Mentoring	65
All Aspects of Industry	66
Aspect of Industry: Planning	67
Aspect of Industry: Management	67
Aspect of Industry: Finance	68
Aspect of Industry: Technical and Production Skills	69
Aspect of Industry: Underlying Principles of Technology	69
Aspect of Industry: Labor and Personnel Issues	70
Aspect of Industry: Health, Safety, and Environmental Issues	73
Aspect of Industry: Community Issues	74
Finding a Job Opening	75
Abbreviations Used in Job Openings	75
Filling Out a Job Application	76
Writing a Letter of Application	76
Sample Application	77
Writing a Résumé	79
Résumé Template	80
Job Interviews	81
Preparing for Job Changes	83
Changing Jobs or Careers	84
Entrepreneurship	85

Successful Employee Characteristics

✔ Communication

Employees need to communicate with supervisors, managers, customers, and coworkers. You must be able to speak, read, and write the language that is used on the job. It is equally important to be able to listen well, to ask questions, and to explain things clearly. Without listening well, you may not know what is expected. Ask for help when you need it. And, if you don't clearly explain information, everyone you work with may look bad.

✔ Honesty

Employers expect their employees to be honest. One lie can destroy your reputation and not only threaten your current job, but future opportunities as well. Would you hire someone if you knew he or she had a history of dishonesty? Avoid the most common ways workers can be dishonest.

- **Time.** Record only the hours you actually work.
- **Money.** In addition to not stealing, accurately account for any company expenses for which you will be reimbursed.

✔ Responsibility

Employers want their employees to accept responsibility for their actions. It can be hard to take responsibility, especially when things go wrong, but it's important to be truthful. Explain what happened, but don't try to blame someone else or make excuses. Do what you can to solve the problem.

✔ Dependability

What one person does, or doesn't do, affects others. If somebody doesn't show up for work on time or leaves early without permission, the other employees may not be able to finish their work. Your employer will not tolerate it and your fellow employees with resent it if they cannot count on you.

✔ Teamwork

A team is a group of people working together to reach a common goal. Even if your workplace is not organized into teams, you still need teamwork skills. One of the main reasons workers lose their jobs is because they cannot get along with others. An employer expects you to work cooperatively with others. A team member

- Plans and makes decisions with others
- Respects the opinions of others
- Realizes there is "give and take" in order to achieve group results
- Encourages and supports fellow team members

✔ Acceptance of Constructive Criticism

Constructive criticism is part of a boss's job. He or she needs to let employees know how they are doing. If your boss tells you that you've done something incorrectly, consider it an opportunity to learn. Ask how you can improve, and follow through on your boss's suggestions.

(Continued on next page)

Successful Employee Characteristics (Continued)

✔ Initiative

Taking initiative means doing what needs to be done without being told to do it. Employers value workers who are willing to go the "extra mile" and who look for opportunities to do more than just what they were hired to do.

✔ Willingness to Learn

Employers want people who can do their job well and follow directions well. Listen carefully to instructions. If you are unsure about how to do something, ask questions. Be willing to learn any job, no matter how small. Learn all you can about your job and about the company. This information will help you do your job better and will prepare you for a promotion.

✔ Respect for Others

Without respect for one another, there can be little cooperation as you work. Since respect is a two-way street, usually the more you give, the more you'll receive. Two attitudes that prevent respect are a message of "I'm better than you" and jealousy. It's important to remember that

- Jealous workers view coworkers as rivals, withhold respect, and make cooperation difficult.
- Each employee—regardless of what his or her job may be—has something to contribute.
- Respecting your coworkers will create a positive, cooperative atmosphere.

✔ Positive Attitude

Your attitude toward your job is a very important factor in your success. You have a choice: you can act positively or negatively toward your job. If you have a positive attitude toward your boss and coworkers, chances are much better that they will respond positively toward you.

A sense of humor also helps. Seeing the light side of a situation can get you–and your coworkers–through stressful times.

This table will help you become more aware of your strengths and weaknesses. Place a check in the column that best describes you. Be honest!

Self Assessment

Trait	Excellent	Good	Fair	Poor	Very Poor
Communication					
Honesty					
Responsibility					
Dependability					
Teamwork					
Constructive Criticism					
Initiative					
Willingness to Learn					
Respect for Others					
Positive Attitude					

Teamwork

Teamwork is a skill you will use in all aspects of your personal life as well as your work life. Working in a team can be fun and challenging. Like any skill, it takes practice.

Companies today rely very heavily on teams of workers to get jobs done. For example, there are design teams, surgical teams, sales teams, and investment teams. On a construction job site, the team is the crew you work with. In each case, the team members must work together to reach a common goal.

Unique Contributions

Each person on a team has something unique to contribute. For example, to design and build a house, there must be people who know architecture, carpentry, electrical work, plumbing, etc. Each contributes to making a house functional and attractive. Besides providing a variety of skills, a team that really works well together can accomplish more than each of the members could on their own.

Are You a Member of a Team?

Working as a team member is probably already a part of your everyday life. Think of the ways you work with other people on a daily basis. For example, at home you are responsible for certain jobs. At school you might be a member of a committee that has to make decisions about a school activity. You might also be on a sports team. Those teamwork experiences have probably helped you prepare for the workplace.

Team Roles

Typically, a team has three or more members. Initially, the role you play on the team might fall into one of the categories shown in the table below. As the project continues, you may find that you play more than one role. For example, a leader is also an encourager and a contributor. All team members are contributors and encouragers.

Team Roles

Role	Functions
Leader	• Keep members focused on the team's goal • Make sure members understand their jobs • Watch the schedule • Set a good example
Contributor	• Help others stay on task • Support other team members • Ask questions • Complete tasks • Evaluate outcomes
Encourager	• Listen • Share ideas • Encourage everyone to participate

(Continued on next page)

Teamwork (Continued)

Teambuilding

As part of a team, you need to focus on two things: completing the task and building and maintaining positive relationships with team members. Building a team that really works well does not happen automatically.

Teams commonly go through four stages before they function at a high level. The first two stages are usually awkward and produce conflict that team members must resolve. They must get to know each other and see how they fit within the team. Some teams may exist for too short a time to reach the highest level of functioning. However, working through the stages will help your team reach the highest level possible.

Stages of Teambuilding

Stage	Tasks	Relationships
Forming	• Set specific goals • Decide what to do and when • Decide who will do what • Decide who will lead or if leadership will be shared	• Get to know each other • See where members fit within the team • Share enthusiasm • Encourage shared participation
Storming	• May use individual ways of getting own parts done • Follow planned process for making all the parts work together	• Make team goals as important as individual goals • Cooperate and get along well; learn to function better • Support, encourage, and guide each other
Norming	• Stay focused on goals • Make decisions • Develop processes for carrying out team plans	• Recognize and accept differences • Develop ways to behave that are acceptable to all • Resolve basic conflicts and build trust • Form a team image
Performing	• Achieve high levels of productivity • Work independently • Take initiative • Focus primarily on getting the work done	• Know how to work together • Achieve a high level of trust • Contribute equally • Resolve conflicts and make decisions quickly • Reach a level of win-win cooperation

Leadership Styles

In a team, you will often play a leadership role. There are several leadership styles that you may use as you complete a task. Each one is useful for varying stages of completing a project.

Democratic Leaders

The democratic leader makes decisions by consulting team members and allowing them to decide how the task will be tackled and who will perform each part. This type of leader

- Makes sure that team members know their jobs and stay focused on the goal

(Continued on next page)

Teamwork (Continued)

- Values group discussion and input from the team
- Draws from a pool of team members' strong points in order to obtain the best performance
- Motivates the team by empowering the members to direct themselves
- Guides them with a loose reign

Coaching Leaders

As a project progresses and teams function at a higher level, the team leader can become more like a coach. This type of leader helps team members stay focused on the goal, watches the schedule, and provides direction and guidance as needed.

Autocratic Leaders

The autocratic leader dominates team members and delegates tasks. Generally, an authoritarian style is not a good approach to getting the best performance from a team at the beginning of a project. It may become necessary towards the end as time runs out. Also, some situations may call for urgent action, and in these cases an autocratic style of leadership may be best. In some situations, team members may actually prefer an autocratic style.

Laissez-Faire Leaders

The laissez-faire leader uses little control, leaving team members to sort out their roles and tackle their work. This approach can leave the team floundering with little direction or motivation. However, it can be very effective when a team of highly skilled, highly motivated people have reached a high level of functioning. Once a leader has established that a team is confident, capable, and motivated, it is often best to step back and let the team get on with the task. Interfering can generate resentment and detract from its effectiveness. By handing over ownership, a leader can empower a team to achieve its goals.

Team Problem Solving

In the workplace you may be assigned to a team. The team is given a problem to solve. Here are some basic steps for solving problems.

1. **State the problem clearly.** Stating the problem clearly in a sentence or two often helps to clarify the problem. Figuring out what the problem really is may bring you halfway to a solution.
2. **Collect information.** Gather as much information as possible that could be used to solve the problem.
3. **Develop possible solutions.** There is usually more than one possible solution to a problem. Brainstorm to come up with as many ideas as possible. Do not judge the ideas. The more ideas the team can think of, the more options your team can consider in the next step.
4. **Select the best solution.** Weigh the advantages and disadvantages of each possible solution. Consider all the factors and base the decision on the team's goals and the particular situation the problem presents.
5. **Implement the solution.** Try out the best choice. Now you will see if the solution is workable.
6. **Evaluate the solution.** This is not the end of the problem-solving process. A good solution should be evaluated carefully to make sure it is appropriate to the problem at hand. If it doesn't work, try another.

Try This Out

Picture yourself on a team at work. Your team has been assigned an important problem to solve.

- How do you think the concepts of teamwork will help you?
- Will this make your job easier or harder? Why?
- Would it change your work habits or your approach to the job?

Apprenticeship Training

Apprenticeships are a popular form of training in carpentry and other trades. Apprentices learn their trade by working alongside highly-skilled workers in a formal training program. These programs combine on-the-job training with related classroom instruction. Apprentices are paid wages that increase as their skills progress. Employers value workers trained through apprenticeships because this type of training requires dedication, development of good job habits, and learning real-world trade skills.

Qualifications

Various programs have different requirements for apprenticeships. Many also test applicants for aptitude for the type of skills the trade requires. Applicants who have completed high school with coursework in carpentry, mechanical drawing, and mathematics are preferred. Work experience in the armed services or Job Corps is also beneficial. Because there are a limited number of apprenticeships, coordinators try to identify the most qualified applicants for their programs.

Registered Apprenticeship Programs

Registered apprenticeship programs are run either by an employer, groups of employers, or by a group of employers who work with a union or association. These groups of people provide the training and wages for the apprentice. They fulfill government requirements set by the United States Department of Labor, Employment and Training Administration.

Depending upon the structure of the employer's business and the other organizations involved in the apprenticeship, the program will either be union or non-union. The apprentice and his employer have a written agreement stating the requirements and responsibilities that both will provide.

Unions have a long history of providing apprenticeship programs. The training of apprentices is overseen by a joint apprenticeship and training committee (JATC), which tests apprentices to make sure they have the proper skills for their trade. Jointly sponsored programs have both labor and management representatives on the committee.

Union carpentry apprentices begin earning about fifty percent of what journey level workers, employees who have completed training, earn. Eventually, apprentices earn up to ninety or ninety-five percent of full wages. Union apprentices also have the security of a union contract, which guarantees good wages, health insurance, a pension, and continuing skill training.

Rewards of Apprenticeship

Participating in an apprenticeship program reaps many benefits. Apprentices receive education paid for by their apprenticeship program, earn wages, and learn the skills employers demand.

Contact Information
- www.doleta.gov
- Contact your local union or building trade association.

Work-Based Learning

In order to provide real work experience, schools and businesses team together to provide students with work-based learning related to a chosen career. The chart below shows types of work-based learning opportunities you may encounter. More detailed information follows.

Type of Program	Description	Purpose
Service Learning	Students participate in an organized work experience that meets the needs of the community.	Apply classroom learning to real-life situations for personal growth and to learn civic responsibility
Job Shadowing	Students follow someone on the job.	Learn about a career through watching and listening
Co-op	Students are employed in a job directly related to their program of study. Schedules vary with time split between school and the job.	Apply classroom learning to work situations
Internship	Students work alongside current workers, gaining experience while completing general courses required for high school graduation.	Supervised practical training, usually near the end of academic preparation
Apprenticeship	Students combine structured, paid work and on-the-job training.	Prepare for professional employment in a skilled occupation
Job Rotation	Workers learn a variety of skills by doing many different jobs at a worksite.	Develop flexibility and competency in a variety of skills
Mentoring	An adult worker or supervisor introduces the student or new employee to a job, a profession, or a skill area.	Model workplace behavior and skills; "mentee" can turn to mentor for feedback, support, and career guidance

Service Learning

Service learning programs combine meaningful community service with academic learning, personal growth, and civic responsibility. In these programs, students apply classroom learning to real-life situations in the community. They are exposed to the world of work in a setting where they experience firsthand the needs of the real world. They experience a setting in which they can improve their communities, learn skills they can use at future worksites, and develop a sense of caring for others.

Service programs provide structured time for students to reflect on what they did and saw during the community service activity. Students often keep a journal throughout the experience.

(Continued on next page)

Work-Based Learning (Continued)

Service learning programs provide opportunities for
- Exploring new roles and interests
- Taking responsibility for and accepting consequences of your own actions
- Seeing how classroom learning is needed for real life
- Acquiring practical skills
- Developing insight, judgment, caring, and understanding
- Acquiring knowledge and exploring service-related careers
- Gaining a chance to relate to people

Job Shadowing

Some employers may provide job shadowing experiences in order to expand awareness of a related field. Job shadowing involves brief student visits to a variety of workplaces, during which time students "shadow," observe, and ask questions of individual workers. Classroom exercises conducted prior to and following the job shadow help students connect their experience to their coursework, career pathways, related skills requirements, and future educational options.

Job shadowing
- Provides a realistic view of a specific job
- Allows students to observe employees on the job
- Allows students time to ask questions

Cooperative Education

Some schools offer cooperative education programs. These are school-based programs that combine classroom study with hands-on experience. Following a carefully developed plan, students work in jobs related to their major field of study while attending school. They can relate or apply what they are learning in class to their major field of study while attending school. Likewise, they can relate the work experiences to what they are learning in class.

Both students and employers benefit from cooperative education programs. While at work, students have the opportunity to develop their technical skills and gain experience in working with other employees and clients. At the same time, they are often paid for their work, which helps to offset their school expenses. Students also receive graded academic credit. Employers benefit because most students have a genuine interest in the field and are motivated to do a good job and learn all they can.

Internships

Many schools offer programs in which students can serve internships at local businesses. An internship provides supervised practical training. Students work alongside current workers, gaining experience while completing the general courses required for high school graduation.

Internships may be paid or unpaid. Internships allow students to observe the world of work and develop work skills targeted to their chosen career field. Students also learn work-related terminology and the culture and expectations of the workforce.

Apprenticeships

Apprenticeship programs are designed to help people develop the skills needed by industries today. They may be sponsored by labor groups, employers, employer associations, or a combination of these. Formal apprenticeship programs typically last three or four years.

(Continued on next page)

Work-Based Learning (Continued)

Apprentices participating in a registered apprenticeship program enter into an agreement with the sponsor. The apprentice agrees to perform the work and complete the related study. The sponsor agrees to make every effort to keep the apprentice employed and to comply with the standards established for the program.

Upon entry into the apprenticeship program, apprentices are paid a progressively increasing schedule of wages. As the apprentices demonstrate satisfactory progress in both the on-the-job training and related instruction, they are advanced in accordance with the wage schedule as outlined in the registered apprenticeship standards.

Job Rotation

Some employers hire entry-level workers and use job rotation. In order for the workers to develop all the necessary skills, the employers have workers rotate through a variety of jobs. By doing so, the workers develop their abilities to learn quickly as well as the various job skills.

Mentoring

Mentors often work with supervisors and classroom teachers to guide students in the work experiences described previously. *Mentors* are usually experienced workers who have the skills and knowledge that the students need to learn to be effective on the job. They might also be other adults such as work-study coordinators, work-experience counselors, or job coaches.

Mentors work one-on-one with each student. Mentors

- Serve as coaches and on-the-job instructors
- Provide students with feedback
- Provide instruction in the culture and expectations of the workplace
- Serve as role models for students as they learn and practice good work behavior, such as on-time performance, dependability, and meeting quality standards
- Provide occupational information related to a particular industry

The skills students learn in these work experiences will prepare them for the continuing education they will receive as part of the workforce. By using these opportunities, students will succeed.

Workplace Skills Checklist

The following checklists will help you identify skills you have and those you need to be successful in the workplace.

Seeking and Applying for Employment Opportunities

- ❑ Locate employment opportunities.
- ❑ Identify job requirements.
- ❑ Identify conditions for employment.
- ❑ Evaluate job opportunities.
- ❑ Prepare résumé.
- ❑ Write job application letter.
- ❑ Complete job application form.
- ❑ Prepare for job interview.
- ❑ Send follow-up letter.

Accepting Employment

- ❑ Complete state and federal tax forms.
- ❑ Complete withholding allowance certificate form (W-4).

Communicating on the Job

- ❑ Communicate clearly with others, including those from other cultures.
- ❑ Ask questions about a task.
- ❑ Read and follow written directions.
- ❑ Prepare written communication, including work orders.
- ❑ Interpret the use of body language.
- ❑ Use good telephone etiquette.
- ❑ Listen to directions and follow them.
- ❑ Use proper e-mail etiquette.
- ❑ Write in legible handwriting.

Demonstrating Teamwork

- ❑ Match team members' skills to group activities.
- ❑ Encourage shared participation.
- ❑ Provide support to team members.
- ❑ Build and maintain trust.
- ❑ Complete team tasks.
- ❑ Evaluate outcomes.

Maintaining Professionalism

- ❑ Treat all people with respect.
- ❑ Exhibit positive behavior.
- ❑ Use job-related terminology.
- ❑ Participate in meetings in a positive and constructive manner.

Maintaining a Safe and Healthy Environment

- ❑ Follow environmental practices and policies.
- ❑ Comply with safety rules and procedures.
- ❑ Use and maintain proper tools and equipment.
- ❑ Maintain the work area.
- ❑ Act appropriately during emergencies.

Demonstrating Work Ethics and Behavior

- ❑ Show up on time.
- ❑ Follow rules, regulations, and policies.
- ❑ Carry out job responsibilities.
- ❑ Assume responsibility for your decisions and actions.
- ❑ Don't abuse drugs or alcohol.
- ❑ Demonstrate willingness to learn.

(Continued on next page)

Workplace Skills Checklist *(Continued)*

Demonstrating Work Ethics and Behavior (Continued)

- ❏ Practice time management.
- ❏ Practice cost-effectiveness.
- ❏ Display initiative.
- ❏ Show pride in your work.

Using Resources

- ❏ Avoid waste and breakage.
- ❏ Plan your time to accomplish tasks.
- ❏ Make a list of supplies and materials needed to do a task.
- ❏ Follow a budget for projects.

Using Information

- ❏ Read instructions and understand how they affect your job.
- ❏ Check supplies or products received against an invoice or packing slip.
- ❏ Find and evaluate information.
- ❏ Use a telephone directory.

Using Systems

- ❏ Understand how your department fits within the whole operation.
- ❏ Find out what work is done in each department and how it fits into the operation.

Using Interpersonal Skills

- ❏ Teach others how to perform a task.
- ❏ Assist customers with problems.
- ❏ Work well with people from different ethnic or cultural backgrounds.
- ❏ Respond to praise or criticism.
- ❏ Provide constructive criticism.
- ❏ Channel and control emotional reactions.
- ❏ Help resolve conflicts.
- ❏ Report sexual harassment.

Demonstrating Technology Literacy

- ❏ Operate and maintain tools and equipment.
- ❏ Use the computer to locate information via the Internet.
- ❏ Use new technologies in your workplace.

Identifying Work Responsibilities

- ❏ Describe responsibilities of employees.
- ❏ Describe responsibilities of employers.

Solving Problems

- ❏ Identify the problem.
- ❏ Use reasoning skills.
- ❏ Assess employer and employee responsibility in solving a problem.
- ❏ Identify solutions to the problem.
- ❏ Select and implement solutions.
- ❏ Evaluate options.
- ❏ Estimate results of implemented options.

Adapting/Coping with Change

- ❏ Exhibit ability to handle stress.
- ❏ Recognize need to change or quit a job.

Working with Employers

Work experience programs such as job shadowing, internships, apprenticeships, and cooperative education programs provide valuable work experience for students. In addition, mentors work with supervisors and classroom teachers to guide students in these work experiences. As students see the connections between school work and the knowledge and skills required by good careers, they understand the importance of learning and can make better decisions about their future.

The extent of work experience and mentoring programs depends on the arrangements made between the school and the employers. With a job shadowing experience, the mentoring relationship will be short. For an apprenticeship, which might last a year or two, the relationship will be more involved and require more planning.

Written Agreements

Once businesses have committed to providing work experiences or mentoring students, agreements should be established. Topics to address include

- What is expected from the employer?
- How will students get to and from the job?
- How many hours will students work per day and per week?
- What is expected regarding attendance?
- How will the job be supervised and how will the student be insured?
- What can the employer expect from the school (numbers of visits by school personnel to the job site during a term; whom to contact in case of a problem or emergency)?

Once both parties have agreed on how these issues will be handled, draw up a written training description that will be signed by the employer, a school official, the student, and the student's parent or guardian. You may wish to use the Work-Based Learning Agreement on pages 63-64.

Aligning Work and School Learning

Once the work-related issues have been resolved, discuss with employers the knowledge and skills the students will be able to take from the classroom to the worksite. If employers feel that skills are missing, instructors may be able to add them to the classroom curriculum. Ask employers to detail the experiences they will provide for each student and the new skills the student will develop. At this time, instructors should provide suggestions for creating an optimal work experience.

It is also important to conduct periodic meetings regarding the effectiveness of the program and the need for occasional adjustments for continued improvement. Teamwork among teachers, mentors, and employers will lead to success for all.

Recruiting and Training Mentors

A good mentoring program will enhance the success of work experiences, and this success depends on recruiting good mentors—sometimes a difficult task. Some people may be shy about volunteering, but may readily agree to serve if they are asked and given guidance by the school personnel and employer setting up the program.

Each semester, or whenever there is another group of employers providing mentors, school site personnel may want to conduct a brief training program. In this program (which is best limited to no more than four hours), instructors and program coordinators can discuss the role of mentors and the school's expectation of the mentoring program.

It is important for instructors and employers to meet periodically to discuss safeguards that protect students from accident or injury and shield the school and the workplace from liability. Because of liability, school personnel must place stringent requirements on student reporting when they leave the campus to go to a worksite.

Liability issues generally fall into two categories: worksite injuries and accident injuries occurring while students are in transit to the worksite or back to school. Therefore, it is important to obtain professional advice and to follow policies based on that advice.

Liability at the Worksite

If a student is considered to be employed (paid or unpaid) under state or federal law, workers' compensation must be obtained by the employer. If the student is paid through an intermediary (such as a youth summer jobs program or a school district) the intermediary must provide workers' compensation.

Liability During Transit

Liability for accidents and injuries during travel from school to the worksite, or from the worksite to school, is the responsibility of the party providing the transportation.

See also:
- Classroom/Lab Safety Checklist, p. 30
- Student Emergency Information Sheet, p. 31

Work-Based Learning Agreement

Student Name _____ Social Security No. _____

Address _____ Telephone _____

Birth Date _____ Work Permit No. _____

Student Career Objective _____

Dates of Employment _____ Hours Per Week _____ Pay _____

Employer _____

Address _____

Supervisor _____ Telephone _____

Student Responsibilities

1. The student will arrive at work on time and will maintain a good attendance record.
2. If unable to report to work for any reason, the student will notify the employer and school coordinator.
3. The student will adhere to company policy; employment may be terminated for the same reasons as regular employees.
4. The student will report job problems to the employer and school coordinator.
5. The student is responsible for his/her own transportation.
6. The student will maintain regular attendance and satisfactory work performance at school; failure to do so will be grounds for dismissal from the job.
7. The student will report to school for designated meetings and instruction related to the work-based learning assignment.
8. The student is not eligible for unemployment compensation; nor may he/she use wages received under this program as credit for unemployment compensation benefits.

School Responsibilities

1. The program is under the direct supervision of the student's instructor and/or a school coordinator.
2. The student will receive related instruction from the school instructor during employment.
3. The instructor and/or school coordinator will visit the student and training supervisor on a regular basis at the training site.
4. The school agrees to award credits earned in the Work-Based Learning Program toward the student's high school diploma.
5. Instructors and/or school coordinators will hold regular meeting with students to discuss progress, issues, and concerns.

WORK-BASED LEARNING AGREEMENT *(Continued)*

Employer Responsibilities

1. The employer will adhere to all state and federal regulations regarding employment, including child labor laws, minimum wage laws, and worker's compensation.
2. The student will be given a variety of work assignments and will be supervised by an experienced employee.
3. The employer will make a periodic evaluation of job progress on a rating form provided by the school.
4. The employer will arrange a conference with the school coordinator when a student problem arises.
5. The employer will provide necessary safety instruction throughout the student's training period.
6. The employer will not employ a student learner to displace a regular worker.
7. The employer is not liable to the unemployment compensation fund for wages paid to the student while under the training program.
8. The employer agrees to follow the training plan developed by the program's school-site instructors and on-the-job supervisors.
9. The employer agrees to notify the school coordinator if the student is absent without notification.
10. The employer agrees to permit the school's representatives to visit the student and supervisor at the place of employment to determine progress, obtain direct feedback, and to make adjustments in the training plan whenever necessary.
11. The employer will hold regular meetings of student learner and supervisor to discuss progress, issues, and concerns.
12. The employer agrees to provide adequate equipment, materials, and other facilities required in order to provide an appropriate learning experience for the student learner.

We, the undersigned, agree to the conditions and statements contained herein.

Student _____ Date _____

Parent/Guardian _____ Date _____

Employer _____ Date _____

Instructor/Coordinator _____ Date _____

Workplace Mentoring

Workplace mentoring programs link students with mentors or work-based coaches, who share their experience with students in the workplace. Students might be in work-based experience programs such as internships or cooperative education programs. Some unions have trained mentors for apprentices.

Mentors and worksite supervisors collaborate with classroom instructors to guide and challenge students. Through work experience, students learn the knowledge and skills appropriate to a specific career as well as the general work expectations of promptness, teamwork, and commitment.

Why Provide Mentors?

Mentoring is a flexible and motivating way to transfer job-related skills and teach positive worker characteristics on a one-to-one basis. A mentor can focus on the specific skills the student needs to acquire. Mentors can also help students establish career goals. Students learn about various kinds of work and see firsthand the school-based skills that they need to acquire to succeed.

What Does a Mentor Do?

A mentor willingly agrees to help a student learn and grow by sharing their skills and experience. To do so, mentors
- Serve as coaches and on-the-job instructors
- Provide focused skill development
- Provide opportunities for observation and hands-on work in a given area
- Provide feedback related to students' work performance
- Provide a sounding board for the student
- Provide instruction in the culture and expectations of the workplace
- Serve as role models by practicing good work behaviors, such as on-time performance, dependability, teamwork, and meeting quality standards
- Consult with classroom teachers and the employer (or supervisor)
- Play an important role in resolving problems that may arise in the workplace
- Provide occupational information related to a particular industry and help explore career options
- Encourage students to continue learning and training for a future career
- Provide other work-related information

People who volunteer to be mentors should
- Be interested in working with young people
- See their roles as an opportunity rather than just another work assignment
- Be willing to share their skills with younger people
- Keep their relationships on a professional level
- Set high standards for those they are mentoring

Because mentoring is a student support system, family members and counselors also play a significant role in the development and support of the program. They should take part in the planning and organization of the program. A team effort is vital to the success of the program.

Benefits of Mentoring

Effective mentoring programs also benefit schools, businesses, and industries. They support good community relations and enhance the quality of the future workforce. Mentoring experiences also enrich and expand opportunities for the mentor. Benefits for the mentors include
- Respect of others in the organization
- Motivation to sharpen their own skills
- Assistance with projects. A student may have technical skills related to a project the mentor is working on.

Supervisors also benefit from mentoring programs. Benefits to supervisors include
- Another source for employee development and problem solving
- Increased motivation of employees
- Better retention of employees

All Aspects of Industry

Maintaining a successful career requires more than technical skills. You will also benefit from experience in all aspects of industry. The phrase "all aspects of industry" refers to the aspects listed in the left column of the table shown below.

Experience with all aspects of industry will help your career in a variety of ways. Throughout life you may need to make career changes. If you are trained to fill only a current job opening, you may be competing for a limited income potential. However, if you have experience in areas such as planning, management, and community issues, you will find more career choices. You can transfer skills like planning and management to different careers or to different responsibilities within the same career. You will also be better prepared to adapt to changes in technology throughout your career.

Choosing academic classes, on-the-job training and work assignments that teach you different aspects of industry will make you a stronger job applicant and employee.

Aspects of Industry

Aspect of Industry	Description
Planning	Planning for the variables that affect the success of a company, such as deciding what products or services to provide and setting goals, policies, and procedures
Management	Methods typically used to manage enterprises; management structure
Finance	Accounting and financial decisions for handling capital
Technical and Production Skills	Skills needed on the job, such as the ability to measure quickly and accurately, read blueprints and drawings, work as part of a team, obtain technical information, and identify requirements for a specific job.
Underlying Principles of Technology	Characteristics of technology used in a specific industry, changing technology, analyzing new equipment for possible use, upgrading job skills, and being adaptable
Labor and Personnel Issues	Worker rights and responsibilities, cultural sensitivity, personnel issues
Community Issues	The impact of industry on the community and the community on industry
Health, Safety, and Environmental Issues	Avoiding job-specific health threats; employer and employee responsibilities for a safe workplace; respecting and protecting the environment

Aspect of Industry: Planning

Many variables affect the success of a company in providing a product or service. To ensure a successful product or service, companies must plan carefully.

Activity: Levels of Planning

Select a level of planning from the table shown to discuss with your worksite supervisor. Find out how the planning takes place, who is involved, and the steps of the process. Prepare a report. Ask your supervisor to review your final report before submitting it to your instructor or presenting it to your class.

Level of Planning	Description
Organizational Level	• Identifying customer wants and expectations • Deciding which products and services to provide, at what levels of price and quality • Anticipating market trend changes • Structuring the organization to effectively support the products and services being provided
Process Level	• Organizing/reorganizing the functions and processes into effective, efficient configurations • Involving management and employees in setting goals, policies, and procedures
Individual Job/ Performer Level	• Setting performance standards • Creating logical work flows without interference from unrelated tasks • Providing adequate resources (for example, time, tools, staff, and information) and training • Providing feedback to employees

Aspect of Industry: Management

Employees work more easily within the company parameters if they understand key management principles. These principles are usually guided by the company's mission statement and the "company culture."

Activity 1: Managing Skills

Divide into teams and research MacGregor's X and Y Management Theory. Discuss the skills and abilities that the modern workforce requires from management.

As a team, present your findings to the class. Engage the class in a discussion about employee and employer relationships. Have the class create a list of skills that a manager should possess.

Activity 2: Functions and Responsibilities

Understanding the functions and responsibilities of management is critical to the success of any worker. Work with your supervisor or mentor to understand the management structure of the company that you work for. Use an appropriate computer software program to create a managerial organizational chart. Then write a narrative describing the duties and responsibilities of each managerial position. Discuss the relationships and interaction between the various managerial positions. Present your findings to the class. Ask your classmates how the companies they work for are similar or different.

Aspect of Industry: Finance

A company's financial and accounting strategies play a large part in its success in the global marketplace. All employees should understand certain basic facts about their company's strategies for obtaining capital. Other facts depend on the type of company. For example, if the company bids on job opportunities, employees should understand how the company estimates costs and figures the bids. In addition, employees should understand the importance of accounting and cost containment in a business, as well as how paycheck deductions affect employees.

Activity 1: Company Financing

The purpose of this activity is to increase your understanding of how a company finances its operations and how it manages cash flow. Work with your worksite supervisor to arrange an interview with the company's comptroller, finance officer, or principal accountant.

Note: Companies may be reluctant to release financial information, so ask your instructor to discuss this assignment with your worksite supervisor before you arrange the interview described in this activity.

Prepare a list of questions before the interview. Questions you might ask include

- What types of credit does the company use to conduct business? If so, for what purpose?
- How does the company manage its cash flow? What role does borrowing play in the management of cash flow?
- Does this company have debt capital? If so, what collateral has been pledged to obtain the debt capital? When does the company plan to be free of debt capital?

Using what you have learned, write a scenario in which you are the financial officer of a company. Describe the business. Explain how the information you learned from the interview would help you keep the company in good financial shape.

Activity 2: Business Loans

Make an appointment to interview a loan officer at a local bank. You may want to do this in teams of two or three. Explain to the loan officer that you are working on a class project and would like information on business loans.

Some of the questions you might ask are
1. What criteria does a bank use to determine whether or not a business is loan-worthy?
2. How do these criteria differ if the request is for a new business or for expansion of an existing business?
3. Does it make a difference whether the business is a sole proprietorship or a corporation? Why?
4. What is the most common reason a bank refuses a business loan?

Prepare a report, using the information you have gathered. Explain how this would help determine what type of business loan to apply for.

Aspect of Industry: Technical and Production Skills

Technical and production skills refer to the skills you need to do construction work. For example, you will need to

- Speak and write clearly
- Use time effectively
- Work in teams
- Read prints and drawings
- Use technical information
- Apply mathematical skills
- Apply construction skills

Activity: Skills for Construction

Contact a contractor and, with the contractor's permission, visit a construction site. Identify the technical and production skills needed for positions related to construction. To guide your research, find answers to the following questions.

- What kinds of structures are built?
- What process does each project go through from beginning to end?
- What skills must key construction workers possess?
- What kinds of drawings or prints must be read and followed?
- What types of math and science are used on the job?
- What types of technical information must be located and used?
- What types of work are done in teams?
- What kinds of interpersonal skills are needed?

After you have completed the research, make a list of the skills needed for positions related to construction.

Aspect of Industry: Underlying Principles of Technology

Technology is the use of knowledge, tools, and resources for practical aspects of life and work. There are many types of technology. Some might be based on mechanical things, while some are based on computers and electronic devices.

Many companies rely on a variety of technologies. Companies can rely on older technology as well as the latest in state-of-the-art technology. A newer technology might change more rapidly.

Activity 1: Current and Future Technology

Find current trade publications for construction or carpentry. Using those and other sources, find out what technology is being used today. What scientific discoveries are affecting technology in that industry? What effect do the new discoveries have on career opportunities? Prepare a report, using the information you gathered in your research.

Activity 2: Technology on the Job

With your worksite supervisor's permission and guidance, perform a technology analysis of the company you work for. Identify the types of technology used in each department. Then determine which types of technology are "cutting edge," as well as which are in line to be replaced and why. What issues must be addressed when analyzing technology equipment for possible use? This report should also project the technology skills employees will need in order to perform their jobs.

Aspect of Industry: Labor and Personnel Issues

Companies and workers have rights and responsibilities that are maintained in various ways. Labor unions represent and defend employee rights. The government has enacted laws to help regulate many personnel issues, as seen in the chart below. Companies have created policies that promote respect among workers and the development of cultural sensitivity. Employers and employees benefit from understanding and using these methods of creating fair and supportive workplaces. Since each company has its own policies and government laws affect many companies differently, it is important to understand labor and personnel issues in general and the issues particular to each company.

Fair Labor Standards Act	The act establishes standards for employment issues such as minimum wages, overtime pay, record-keeping, and child labor.
Civil Rights Act	Employers may not discriminate on the basis of race, color, national origin, sex, or religion. The act also protects U.S. citizens working for U.S. companies overseas. The act also deals with unlawful harassment in the workplace.
Equal Employment Opportunities Act	This is an extension of the Civil Rights Act. It requires businesses to have affirmative action programs for locating, hiring, training, and promoting women and people of color.
The Age Discrimination in Employment Act	People over 40 years of age and older are protected from being discriminated against in any aspect of employment.
Americans with Disabilities Act	The purpose of this act is to prevent employers from refusing to hire or promote qualified disabled persons and to ensure that all employees are treated equally. It requires that public facilities make "reasonable accommodations" for the disabled.
Immigration Reform and Control Act	Only U.S. citizens and people who are authorized to work in the U.S. may be legally hired.
Immigration and Nationality Act	This act is the basic immigration law for the United States. It is used to decide who can enter, how long they can stay, and when they must leave. It also determines who can become a naturalized citizen.
Federal Employment Compensation Act	This act protects employees who are injured or disabled due to work-related accidents.
Occupational Safety and Health Act (OSHA)	This act promotes safe and healthful working conditions.
National Labor Relations Act (NLRA)	This act gives employees the right to organize into unions.

(Continued on next page)

Aspect of Industry: Labor and Personnel Issues (Continued)

Activity 1: Finding More Information About Labor Laws

Labor laws (listed in chart) affect everyone who works in business and industry. To learn more about them, divide into teams of three to five students and research one of the laws and the federal agency responsible for enforcing it. You can find information in the library, through government agencies, or on the World Wide Web. Create a presentation in which you

- Identify major provisions of the law
- Tell what year the law was enacted and whether it was revised
- Discuss how the provisions of the law protect workers
- Describe the implications for employers
- Describe how the law affects workers in your industry

Each team should make its presentation to the entire class.

Activity 2: Labor Unions

The National Labor Relations Act (NLRA) gave workers the right to organize into unions. A union is a group of employees united to bargain for job improvements. Unions use various methods to negotiate with companies. They may use *collective bargaining* to try to reach an agreement. If the union and the company cannot agree, union members may go on a *strike,* during which employees stop working in order to try to force the company to agree to the union's demands. The process should result in a contract which both the union and the company can agree upon.

Use the Internet to research a union such as the United Brotherhood of Carpenters, the United Association of Plumbers and Steamfitters, or the Cement Masons and Plasterers Union.

- How does the union operate?
- Who can be a member of the union?
- What kinds of improvements does the union support?
- How has the union impacted the eight-hour workday and 40-hour workweek?
- What kinds of negotiation strategies does the union use?
- What kinds of dues do the members pay?

Record your findings and report them to the class in a short oral presentation.

Activity 3: Job Environments

Many aspects of the workplace affect a person's effectiveness on the job. One factor that motivates workers to perform at their best is the rewards received for improving and maintaining job performance. Rewards can be both *external* (from without) and *internal* (from within).

Once the basic level of external reward is reached that will satisfy employees, internal factors have been shown to be even more important. Workers will perform in a satisfactory manner when the external factors are present. However, internal motivators must be present as well for peak performance.

(Continued on next page)

Carpentry & Building Construction Instructor Resource Guide
Copyright © Glencoe/McGraw-Hill

Aspect of Industry: Labor and Personnel Issues (Continued)

Examples of external factors include
- Wages or salary
- Working conditions
- Time off with pay
- Health insurance
- Retirement plans
- Recognition for achievements
- Positive feedback

Examples of internal factors include
- A sense of accomplishment
- An appropriate variety of tasks
- Mutually agreed upon goals
- Responsibility
- Interpersonal relationships
- Growth and advancement
- Success based on one's hard work and ability

Internal factors vary according to individuals. For example, people who are high achievers like to have clearly defined goals, personal responsibility for their success, and control over available resources. In contrast, other people place a higher value on being with people than on goals. They like to work in groups and call on friends when they need help.

Think about the environment in which you work.
- How many of the external factors are present?
- How many of the internal factors are present?
- How do you feel about your job? Do you like it? Do you feel appreciated? Do you enjoy interactions with others on the job?

Divide into groups and compare answers. Choose a spokesperson for your group to give a summary of the discussion.

Activity 4: Comparing Personnel Policies

Most companies that employ 50 or more people develop policies to govern their behavior. Examine the personnel policies of two companies, comparing the types and the complexity of the policies. What rights and responsibilities do workers have when companies develop policies? How do the policies differ? Why might two companies have different personnel policies? How might the particular choices in governing employee behavior suit the particular nature of the company? Share your findings with the class in a short presentation.

Activity 5: Cultural Diversity

As business becomes more global, it is becoming more important for us to understand other cultures. We need to do our best to understand and respect cultural differences.

Research a country you are interested in. Imagine that there is a worker shortage in that country, which has provided you with the opportunity to work there for a year. Describe the ways you might need to act differently while in that country. As an alternative, imagine that you will be working with someone who has come from that country to live and work here for a year. What kinds of differences might there be in the way you communicate? How could you overcome these difficulties? Where could you obtain resources about cultural differences? Report on how you could increase communication and demonstrate understanding with your coworker.

Aspect of Industry: Health, Safety, and Environmental Issues

Responsibility for employee health and safety in the workplace is shared between employer and employees. The company is responsible for providing a safe workplace and for complying with safety and environmental regulations. However, individual employees must take responsibility for many of the details. They should follow company policy to abide by federal regulations. For their personal health and safety, employees must be able to read and understand a Material Safety Data Sheet (MSDS).

Activity 1: Meeting Safety and Health Standards

Research local companies to identify one that has a positive record in dealing with health, safety, and environmental issues. Contact the person in charge of worker safety or community relations. It might be someone in the personnel department. Ask permission to do a 20-30 minute interview for a class project. If you are employed, consider interviewing your worksite supervisor or the person in charge of worker safety. Prepare a report on the health and safety policies that are in effect within the company. Questions you might ask include

- What situations on the job might affect a worker's health or safety?
- What health and safety policies and practices are in effect in the company?
- How might the work of the industry affect the health, safety, and environment of the community?
- What steps does the company take to minimize risk to the community?

Activity 2: Respecting the Environment

Find out from your worksite supervisor what environmental regulations your company must follow. Research trends in handling environmental problems for their industries and what new environmental problems loom on the horizon. Prepare written reports and give a synopsis of your findings to the entire class. If possible, use presentation software to present the highlights of the report.

Afterwards, ask your classmates what they see as the major environmental issues that United States business and industry must address during the next decade.

Activity 3: Jobs and the Environment

Find information about jobs that are related to the environment in some way, for example, waste management and the recycling of environmentally harmful chemicals. For sources, you might try looking on the Internet or using computerized career information delivery systems. Identify ten jobs that might develop as a result of the need to solve environmental problems. Present your findings to the class.

Aspect of Industry: Community Issues

All companies exist within communities. Therefore, it is crucial that companies and their workers interact positively with their community. An understanding of this relationship will benefit the workers, company, and community.

Activity 1: Companies Help Communities

A community benefits from a company located in its area. Companies
- Provide jobs
- Pay taxes, which helps pay for roads and other public services
- Provide incomes that people spend at other companies, such as grocery stores and restaurants
- May attract related industries to the area

Think about your job and how it might benefit your community. Ask your worksite supervisor or other managers for their ideas on how the company has benefited the community. What other related companies exist in your community? What new ones might it attract? What other ways has your company benefited the community? Present your findings to your class.

Activity 2: Communities Help Companies

A company depends on the community to provide many of the resources it needs to do business. For example, it would be difficult for a company to generate all its electricity or to build and maintain all the roads needed to receive materials from its suppliers.

A company must look at several factors when it decides where to locate. Some of the factors might be

- Access to transportation, such as roads and airports
- Access to a skilled workforce
- Access to water, electricity, sewage treatment, and fire protection
- Organizations to support workers and their families, such as schools and hospitals

List ways your community supports the company you work for. Share your list with your class.

Activity 3: Charitable Organizations

By providing assistance to a charitable organization, a person helps the community. The person also has the chance to improve existing skills or develop new skills that may be of benefit in a future career.

In order to help people in need, many charitable organizations need people with specialized skills. For example, they may need carpenters, painters, network specialists, or electricians. One such charitable organization is Habitat for Humanity, which builds homes for people with low incomes. The volunteers and future home owners work together to build the housing. Volunteers needed include carpenters, electricians, and a variety of people who understand the various tasks needed to build a home.

List the skills you are learning on the job or in your school program and how they could be used to benefit your community. How could you personally benefit by donating your time and skills? Share your findings with the class.

Finding a Job Opening

There are many different ways to locate jobs that you are interested in. Listed below are methods that are often used.

- **Personal contacts.** Network with friends, relatives, and acquaintances to find out about possible job opportunities.
- **Direct contact with employers.** Companies often have unadvertised openings available. By getting in direct contact with possible employers, you are telling them that you have initiative, a desirable quality in an employee.
- **The Internet.** Many Web sites are available to help employees find suitable employers, and vice versa.
- **Professional associations.** These groups often maintain areas on their Web sites where companies can post job openings and members can post résumés.
- **Job counseling.** Many schools offer job counseling as well as provide a list of job openings.
- **Help-wanted ads.** Watch the ads in newspapers, professional journals, and other sources. If you want to move out of the area, read newspapers from areas of interest. Many newspapers have online components that often include job openings.
- **Employment agencies.** Both public and private employment agencies help people find jobs. These agencies have listings of available jobs, many of which may not be advertised in a newspaper. The agencies ask information about job seekers' skills, interests, and other qualifications. Then they try to match the job seekers with the jobs on their lists.
- **One-Stop centers.** The one-stop system is coordinated by the U.S. Department of Labor. It connects employment, education, and training services into a network of resources at the local, state, and national level.

Abbreviations Used in Job Openings

Abbreviation	Meaning
appl.	applicants, applications
avail.	available
appt.	appointment
asst.	assistant
co.	company
comm.	commission
const.	construction
dept.	department
eves.	evenings
exp.	experience
ft	full-time

Abbreviation	Meaning
hr.	hour
immed.	immediate
incl.	included
M-F	Monday through Friday
max.	maximum
mfg.	manufacturing
mgr.	manager
min.	minimum
mo.	month
nec.	necessary
pt	part-time

Abbreviation	Meaning
pos.	position
pref.	prefer, preferred
ref.	references
req.	require, required
sal.	salary
tel.	telephone
w/	with
wk.	week, work
yr., yrs.	year, years

Filling Out a Job Application Form

In order to choose the best person for the job, employers need information about the applicants. When you apply for a job, most employers ask you to fill out a job application form. On the form you are asked to provide information in a certain order.

Your instructor will give you an example of an application form. Here are a few tips for filling them out. Following these tips will increase your chances of getting the job you desire.

- **Read and follow the instructions.** For example, does the application form tell you to print, write, or type the information? Does it call for your last name to come first?
- **Answer all the questions on the application.** If a question or section does not apply to you, put NA, for "not applicable." This will show the employer that you did not overlook the question and that you have nothing to hide.
- **Fill out the form neatly.** You might want to photocopy the form and fill it in for a "trial run." Then you can follow it to fill in the final form more neatly.
- **Be as specific as you can possibly be in all areas.** Always be truthful. Lying on an application is illegal.
- **Reread your application before you hand it in.** Check for spelling errors. Be sure all phone numbers and addresses are correct.

Writing a Letter of Application

When you apply for a job by mail or e-mail, the cover letter that accompanies your résumé (or application) is often as important as the résumé itself. The cover letter introduces you and sparks interest in reading your résumé. It's like an advertisement for *you*.

A cover letter should not be longer than one page. It should be friendly but businesslike. Also, it must be clearly written and contain no errors in spelling or grammar.

Following is a general outline of how a successful cover letter can be organized.

First Paragraph: Introduction

The introductory paragraph establishes the tone of the letter and captures the potential employer's attention. It should include the following:

- Your source of information about the position
- The position in which you are interested
- A statement expressing your desire to be considered for the position

These elements do not have to be in any particular order or covered in a single sentence. The more important—and more difficult—goal is to establish a tone of modest self-confidence.

Middle Paragraphs: Experience and Education

The middle paragraphs describe your experience and/or educational background. Start with the stronger area. For example, if your work experience is more pertinent to the position you want, discuss it first.

Do not restate the contents of your résumé. Choose two or three points that will most interest the employer and develop those points. Then refer the employer to your résumé for more details.

Final Paragraph: Conclusion

The goal of the concluding paragraph is to obtain an interview. If you have not already referred to your résumé, do so here. Explain why you want to work for the company. Conclude with a polite, confident request for an interview at the employer's convenience.

Carpentry & Building Construction Instructor Resource Guide
Copyright © Glencoe/McGraw-Hill

Sample Application

Application for Employment

Equal Opportunity Employer

Personal Information

Please print clearly.

Full Name: _____
                  First   Middle   Last

Present Address: _____
                Number and Street   City   State   Zip

Social Security Number: _____ Phone Number: _____

Are you 18 years or older? ❑ Yes ❑ No If not, what is your date of birth? _____

Have you ever been convicted of a felony or misdemeanor? ❑ Yes ❑ No
If yes, please explain in the space below. *(A conviction record will not necessarily preclude employment.)*

_____

_____

Are you aware of any reason that you cannot perform essential functions of the job with or without reasonable accommodations? ❑ Yes ❑ No
If yes, please specify: _____

Availability

Total hours available per week: _____ Position or shift applied for: _____
Please indicate the time you are available for work each day.

DAY	Su	M	Tu	W	Th	F	Sa
FROM							
TO							

Education

SCHOOL TYPE	SCHOOL NAME/ADDRESS	CITY/STATE
Senior High School		
College/University		
Graduate School		
Trade/Business/Night Courses		
Other		

(Continued on back)

Sample Application *(Continued)*

Work History

List most recent jobs first.

Company: _____ From: _____
Address: _____ To: Month   Year
City/State: _____ Phone: _____ _____
Position: _____ Salary: _____ Month   Year
List of duties: _____ Supervisor: _____
Reason for leaving: _____

Company: _____ From: _____
Address: _____ To: Month   Year
City/State: _____ Phone: _____ _____
Position: _____ Salary: _____ Month   Year
List of duties: _____ Supervisor: _____
Reason for leaving: _____

Company: _____ From: _____
Address: _____ To: Month   Year
City/State: _____ Phone: _____ _____
Position: _____ Salary: _____ Month   Year
List of duties: _____ Supervisor: _____
Reason for leaving: _____

To comply with the Federal Employment Eligibility Verification Law, you must bring **EITHER** one document from List A **OR** one document from List B and one document from List C below:

List A
- United States Passport
- Certificate of United States Citizenship
- Certificate of Naturalization
- Unexpired Foreign Passport with Employment Authorization
- Alien Registration Card with Photograph

List B
- Unexpired State-issued Drivers License
- Unexpired State-issued Identification Card
- School Identification Card with Photograph
- Voter's Registration Card
- United States Military Card

List C
- Social Security Number Card (Original)
- An Original or Certified Copy of a Birth Certificate Issued by a State or County, Bearing an Official Seal

---

Fair Credit Reporting Act and Employment At Will Disclosure

I understand that I am applying for employment "at will." If I am hired, employment can be terminated at any time for any reason with or without prior notice or cause. Any oral statements or promises to the contrary are not binding upon the employer.

I confirm that all my answers to the questions in this employment application are accurate and complete. I also understand that the submission of any false information in connection with this employment application may be cause for immediate termination at any time thereafter should I be employed by this company. I understand that my employment will be contingent upon the accuracy, completeness, and acceptability of the information furnished by me. I acknowledge that this company has permission to verify all statements in this employment application.

I certify that the information on this application is accurate and complete.

Signature _____ Date _____

Writing a Résumé

Employers need information about you in order to determine whether you are qualified to work for them. You can provide this information by preparing a résumé. A *résumé* is basically a "sales-catalog" description of yourself. It tells who you are and what you can do. Your teacher will give you an example of a résumé.

Elements of a Resume

A résumé should include the following:

- Your full name
- Your address
- Your telephone number (including your area code) and e-mail address
- A brief phrase or sentence describing your career goals or the job you want
- Your work experience. For each job you list, identify the position you held. Describe your job responsibilities and duties. Include the dates you started and left (if applicable). If you are currently employed, write "Present" in place of the year.
- Your education. List the degree abbreviation and major. List the school by its full name and the month and year you graduated.
- Special skills and abilities. Include anything that indicates you would be good for the job, such as special training, aptitudes, and other relevant information.
- Special honors or activities related to the job you want

There are many other types of résumés. Information is available in reference materials (including computer software) to help you write one.

Action Verbs

When you describe your positions, use action verbs. For example, rather than writing "Was responsible for stocking shelves," write "Stocked shelves" or "Supervised three workers" rather than "Three workers were supervised by me." In thinking about your functions and responsibilities, identify the action verbs that clearly describe what you do or did.

Some examples of action verbs are

- Advised
- Analyzed
- Assembled
- Built
- Completed
- Coordinated
- Created
- Delivered
- Developed
- Devised
- Directed
- Improved
- Increased
- Maintained
- Managed
- Obtained
- Operated
- Organized
- Performed
- Produced
- Researched
- Supervised
- Trained
- Wrote

Résumé Template

Your Name

Street Address
City, State, Zip Code
Telephone Number
E-mail Address

Career Objective   A brief phrase or sentence describing your career goals or the job you want.

Work Experience

Company Name, City, State   Month Year—Present
Job Title or Function. Describe the types of things you do, such as stock shelves, wait on customers, and repair electronic equipment.

Company Name, City, State   Month Year—Month Year
Job Title or Function. Describe the types of things you did, such as stocked shelves, waited on customers, and assisted with carry-outs.

Education

Degree, Major
School Name, City, State
Graduated Month Year
(optional) 0.0 Grade Point Average

Degree Major
School Name, City, State
Graduated Month Year
(optional) 0.0 Grade Point Average or Honors Received

Special Skills/Abilities   Special training, aptitudes, and other relevant information.

Honors/Activities   Club or Activity Name; Title, Year-Year
Club or Activity Name; Title, Year-Year
Club or Activity Name; Title, Year-Year
Club or Activity Name; Title, Year-Year

References   Available upon request.

Job Interviews

Practicing for an Interview

Following are some questions commonly asked during job interviews. It's not likely that you'll be asked all of these questions in one interview. However, if you think about how you would answer them, you will be prepared with a good answer for any that you are asked. Then, set up a mock interview. Have a friend or classmate interview you, using the questions from this list. If possible, do this in front of others. Have them look for ways in which you can improve your performance.

- Why do you want to work for this company?
- How did you find out about this job opening?
- Why do you think you are suited for this job?
- Tell me about your education or training.
- Which courses in school did you like best? Which did you do best in?
- What are your strengths and weaknesses?
- What jobs have you had in the past? Why did you leave them?
- Would you rather work alone or with others?
- What kinds of work do you like to do?
- What type of work do you want to be doing five years from now?
- What salary do you expect to receive?
- What questions would you like to ask me?

Job Interview Do's and Don'ts

Do	Don't
Arrive early	Be late
Dress appropriately	Dress sloppily
Have a firm handshake	Have a limp handshake
Provide a neat, concise, and complete résumé or application	Arrive unprepared
Maintain eye contact	Glance around the room
Use your best manners and smile	Forget to be polite and respectful
Use good posture and sit quietly	Fidget, chew gum, or slouch
Speak clearly and use good grammar	Mumble one-word replies or use slang or profanity
Answer questions clearly, completely, and in a positive way	Try to fake answers, complain, or talk about your problems
Show enthusiasm and confidence	Show a lack of interest
Thank the interviewer and leave as soon as the interview is over	Keep the interviewer from other duties

(Continued on next page)

Job Interviews (Continued)

Being Interviewed

For some people, an interview can be the most difficult part of getting a job. There are many books that give tips on how an interview should be handled. Below are a few helpful interview tips.

- Dress appropriately and be neatly groomed. You always want to look your best. First impressions are important.
- Make sure you come prepared with anything the employer might want, such as references or a résumé.
- Plan to arrive in the office five to seven minutes early. This way, you're sure to be on time.
- Greet your interviewer with a firm handshake and pronounce his or her name correctly. (If you're not sure, ask politely.)
- Wait until you are asked to be seated.
- Don't put anything on the interviewer's desk unless it is asked for.
- Use your body language to show self-confidence. Don't fidget or make nervous movements. Sit leaning slightly forward, and look directly at the interviewer as you listen and as you answer questions.
- Be prepared to answer questions about school or previous employment. The two questions most often asked are: "What kind of work would you like to do?" and "What do you expect to be paid?"
- Be enthusiastic. Show you are interested in the business.
- Don't say anything negative about a former employer. If you were treated badly or working conditions seemed intolerable, just say something like, "I would rather work under different conditions."
- When your interview is over, shake hands and ask for a business card. Be sure to leave with the same self-assurance you had when you came in.
- Follow up with a thank-you note expressing your appreciation for the chance to interview for the job.

Keep in mind that an interview is a time to put your best foot forward and to convince the interviewer that you are the best person for the job. You can be your own best salesperson.

Preparing for Job Changes

Today's workplace offers challenges that earlier generations of workers did not face. Skills needed can change quickly. Employers want *lifelong learners*, people who continue to develop new skills and good employee qualities. By becoming a lifelong learner, you will benefit from job changes.

Keeping Up

Many of today's companies invest heavily in employee training and education for this purpose. At your job, make use of all opportunities to keep your skills and knowledge up-to-date.

When new technology appears in your workplace, get involved right away. Volunteer for tasks that will give you hands-on training.

Growing in Your Job

Continuing to update and improve your skills and knowledge will make you valuable to your employer. It may also help you earn a promotion. A promotion is a job advancement to a position of greater responsibility and authority. Promotions will offer new challenges. Usually promotions also bring increased income.

The people who earn promotions are those who have shown their supervisors that they can handle additional responsibility and authority. What qualities and behaviors do employers look for?

- **Seniority.** Seniority is the position you achieve by working for an employer for a sustained length of time. Greater seniority is usually thought of as indicating greater experience.
- **Knowledge and competence.** Employers want workers who know how to do their jobs, even if the new job requires different skills. Employers also look for workers who go a step beyond this–workers who excel.
- **Willingness to learn.** Employers promote workers who show they want to increase their knowledge and skills.
- **Initiative.** You'll probably advance in your career if you make it clear to your supervisor that advancement is an important goal for you.
- **Perseverance.** Perseverance is the quality of finishing what you start. Employers want to know that you will see a job through to completion.
- **Cooperation.** Employers want people who can get along well with others.
- **Thinking skills.** When considering whom to promote, employers look for people who can think through situations and solve problems.
- **Adaptability.** Employers want workers who can adapt to new situations and get a job done.
- **Education and training.** Employers promote people who have the needed skills and education.

Handling Your New Responsibilities

Getting a promotion may change your work life in many ways. Often it means you'll become a supervisor. Then you will be responsible for both your own work and the work of others.

Be aware that as a supervisor your relationships with your former coworkers will change. You'll be the boss. You must oversee their work and give directions. You will review their performance.

At every level of your career, you make choices which affect your job security and your advancement. Preparing ahead for job changes through education and good work habits will help you meet the challenges you will face in today's workforce.

Changing Jobs or Careers

People can expect to change employers several–perhaps many–more times before they retire. This means that your career and your job security are in your own hands. That can be to your advantage–if you think about and plan for the consequences of your job choices.

Considering a Change

Sometimes you may choose the change because you want to seek new opportunities. At other times, you may be forced to change by events beyond your control.

When you are unhappy at work, a job change may be one solution. Evaluate your situation. Before giving up on your job, consider whether there might be a way to stay and solve the problem. Ask advice from an experienced coworker or friend. If you decide to stay, set work goals for yourself. What do you want to achieve in your job–a pay raise, a promotion, new responsibilities or challenges? How long will you give yourself to reach your goals?

Changing your job should never be a snap decision. Before making a change, you should analyze what is missing from your current job and what you want from a new one. Always make sure you have thought through the change carefully.

Making the Change

The job market changes quickly. Tracking employment trends is one way to prepare for the future. If you decide to change jobs, consider the following:

- What are your work skills?
- Which tasks do you enjoy most? Which do you enjoy the least?
- Consider how you can use your skills in new ways.
- List jobs or careers that interest you. What are the pros and cons of each? How can your current skills and interests be applied to them?
- Use your network, your personal connections, to help achieve your career goals. Many people you know may be aware of others to contact or job opportunities you can explore.
- Don't burn bridges with your present employer. You don't want to lose your current job until you've landed a new one.
- When you have found a new job, give proper notice to your current employer. Most people prepare a formal letter of resignation.
- Don't tell coworkers about your job hunt. Wait to inform coworkers about the new job until after you have given notice.

Entrepreneurship

Starting a Business

Entrepreneurs start, develop, and then run their own businesses. They may start their business venture in one of four ways. For carpentry and building construction, the first three ways are most common.

1. **Starting a new business.** Some entrepreneurs start their businesses from scratch. They see a need and start a business to meet that need. There are many challenges to starting a new business. Entrepreneurs must find a location, buy or build a facility, purchase equipment, hire employees, and find ways to attract and keep customers. They must obtain financing and develop a business plan. A business plan identifies the goods or services that will be offered, tells where the business will be located, outlines the owner's goals, and describes customers and the type of marketing that will be done. The advantage of starting a new business is that the owner can decide just how the business will be developed and run.

2. **Buying an existing business.** People who buy an existing business get the facility, equipment, and even employees. However, it's important to find answers to the following questions before buying. Why is the business being sold? Was it making a profit? Does it have a good reputation? Are the current employees skilled and motivated? What is the condition of the building, the equipment, and the inventory?

3. **Taking over a family business.** Entrepreneurs who take over a family business usually have the advantage of years of experience with the business. However, they must consider the same issues as those who are buying an existing business. In addition, they need to consider possible concerns and conflicts with other family members.

4. **Buying a franchise.** Many businesses are franchises. The entrepreneur who buys a franchise obtains the right to sell a company's products or services within a given area, or territory. The company provides the location, management training, and help with advertising and employee recruiting. In return, the franchise operator pays the company a share of the sales, in addition to the initial purchase price, and agrees to operate the business according to the company's policies.

Characteristics of an Entrepreneur

It takes a special kind of person to be a successful entrepreneur. An entrepreneur is

- **Persistent.** Entrepreneurs work until the job is done, no matter what. They know their livelihood depends on getting the job done and reaching the goals they have set for themselves.

- **Risk-taking.** A good entrepreneur does a lot of research and planning before starting a business. Still, in the final analysis, the entrepreneur must put his or her money and reputation on the line.

- **Responsible.** Successful entrepreneurs take responsibility for their actions. They know they are accountable to their customers, employees, and investors. They keep their promises and treat people honestly and fairly.

- **Creative.** Entrepreneurs recognize opportunities and are always looking for ways to improve their businesses. They may develop new products or find new markets for their products.

- **Self-confident.** If you don't believe in yourself, how can you expect your investors, customers, and employees to believe in you? Entrepreneurs must be confident in their business actions.

(Continued on next page)

Entrepreneurship (Continued)

- **Independent.** Entrepreneurs make their own decisions. They run their businesses the way they believe is best. They will ask for advice, but they make the final decisions themselves.
- **Goal-oriented.** Entrepreneurs set goals and then "go for it." They know what they want and work hard to achieve it. They are driven to reach their goals.
- **Competitive.** Entrepreneurs are always looking for ways to make their product or service better than the competition's. They learn as much as possible about things that might affect their businesses.
- **Demanding.** Entrepreneurs expect a lot from their employees, but they expect even more from themselves. Entrepreneurs need to focus on all areas of their businesses. In order to do that, they expect a great deal from everyone involved in the business, including themselves.

Could You Be an Entrepreneur?

Read each statement in the table below. If a statement strongly describes you, rate it a #5. If it doesn't describe you at all, rate it a #1. If the statement partly describes you, rate it as a #2, #3, or #4. Total your answers and divide by 10. The closer your total score is to 5, the more likely it is that you would enjoy being an entrepreneur.

Rate Your Entrepreneurship Qualities

Qualities	1	2	3	4	5
I am creative.					
I take responsibility for my actions.					
I am independent and like to make my own decisions.					
I am persistent and finish a task, despite difficulties.					
I set goals and try to reach them.					
I like to work at my own pace.					
I believe in myself and what I'm doing.					
I like challenges and am willing to take risks.					
I set high standards for myself.					
I am willing to learn in order to make wise decisions.					

Total Score _____ ÷ 10 = _____

Contents

E-mail Etiquette 88
Basic Computer Skills 89
Internet Permission Contract 91
Internet Evaluation Sheet 92
Using the Internet 93

E-Mail Etiquette

E-mail allows quick, direct communication with people in the next room and around the world. However, there are rules of etiquette to follow when using e-mail. Observing e-mail manners can mean the difference between success and failure in getting your messages read and understood.

Send the Right Message

When sending e-mail messages, you can't share facial expressions, gestures, or voice inflections the way you do when you're face-to-face with someone. It's easier to take a person's intentions the wrong way. Here are some things to keep in mind to prevent misunderstandings via e-mail.

- The person you e-mail may not share your values, cultural background, or opinions. Before you send the message, consider how it will be received.
- Don't write anything in e-mail that you wouldn't write on a postcard. E-mail is not entirely secure and can be intercepted by others. A message can also be saved or forwarded to anyone.
- Don't use e-mail to let off steam. Cool off and compose yourself before sending it.
- Would you say the same thing if the person were in front of you? If not, rewrite it.
- DON'T SHOUT! It's all right to emphasize a word or line in capitals, but use the CAPS lock button sparingly.
- Keep harassment and discrimination policies in mind. Don't write, send, or forward e-mail that is offensive, obscene, discriminatory, or sexist.
- E-mail is less formal than letter writing, so it's tempting to relax on the formalities of grammar and spelling. However, you will be judged by the quality of your writing. Like it or not, spelling and grammar do count.
- Make sure your notes are clear, logical, and concise. It is possible to write a paragraph that contains no errors in grammar or spelling, but still makes no sense.
- Bad information spreads easily on the Internet. Once e-mail is sent, you lose control of where it might be forwarded. Before you send information, check the facts.
- Having good manners yourself doesn't give you the right to correct everyone else. If you decide to inform someone of an e-mail mistake, point it out politely and privately.

Respect Your Reader's Time

- Send messages only to those who really need the information. Mail to a group list only if it's appropriate for everyone on the list to receive the message.
- If you want your e-mail read, use a specific subject line.
- Recipients of your e-mail may not appreciate an inbox filled with recipes, jokes, inspiring stories, or requests for charitable donations.
- Don't use the school or organization's computers to send or forward electronic junk mail advertisements (spam).
- For urgent messages, try another medium. E-mail is quick and efficient, but your reader is not obligated to check or reply to it on a regular basis. If you have a pressing issue, try the phone or arrange a face-to-face meeting.
- When you return e-mail, don't reply to everyone the e-mail has been sent to, unless it's necessary. Delete the e-mail addresses of people who don't need to receive a reply.
- When replying, use the automatic quote feature and edit the quote. Quote just enough of the previous message so that the receiver can see at a glance what has gone on before in the conversation.
- Don't forward e-mail virus warnings; almost all are hoaxes. Antivirus software is the best defense against viruses.

Basic Computer Skills

Many workplaces are using computer technology for various purposes. Having a basic understanding of computers will help you to be a more effective employee.

Three basic operations of a computer are input, processing, and output. Input is received through input devices such as keyboards, mice, or trackballs. Processing is handled by the central processing unit (CPU) and other internal electronic devices. Output is produced by the monitor, printer, and other devices.

The computer's operating system controls the overall activity of the computer. For example, the operating system starts and runs application software, such as estimating programs. The operating system also enables the different components of the computer system (such as the disk drives, printer, and monitor) to work together.

Most business software runs on computers that use the Microsoft Windows operating system. This article describes basic operations of Windows 95, 98, and 2000.

Starting and Shutting Down

In most computers, Windows starts automatically when you turn on the computer. When startup is complete, the screen will display the Windows desktop. The exact look of the desktop varies depending on how your computer is set up, but will probably have the following items.

- **Start button.** Selecting this button displays a menu of items, such as programs and documents. Selecting an item with an arrow next to it displays another menu. At the bottom of the Start menu is the Shut Down option.
- **Shortcut icons.** There will be one or more icons on the desktop. These are shortcuts, and they provide an easy way to open programs or documents.
- **Taskbar.** The taskbar is usually in the bottom area of the desktop. It shows which Windows features are in use and displays the name of any open programs (also called *applications*). You can have more than one program open at a time.

Files and Folders

You will need to save each document or drawing you create as a file. The computer itself also has many different kinds of files it uses to operate. The files are organized into folders, much like paper documents are organized into folders in a filing cabinet. (Folders are sometimes referred to as *directories*.) Each folder is labeled for easy identification. Here are two ways to see the files and folders stored.

- **My Computer icon.** If you select this icon on the desktop, a window displays the contents of your computer. You will see icons for the storage devices used by your computer. This may include a hard disk, compact disc, network drives, or removable storage devices. Selecting one of the icons will display the folders and files stored there. You can also search for and open files and folders.
- **Windows Explorer.** This shows the hierarchical structure of files, folders, and storage systems or devices used by your computer. If you select a folder, Explorer will display its contents. You can copy, move, rename, delete, and search for files and folders. You can also create new folders.

Deleting Files

When you delete files or folders from your hard disk, Windows places them in the Recycle Bin, where you can retrieve them later if you need to. Files or folders deleted from a floppy disk or a network drive are permanently deleted and are not sent to the Recycle Bin.

(Continued on next page)

Carpentry & Building Construction Instructor Resource Guide
Copyright © Glencoe/McGraw-Hill

Basic Computer Skills (Continued)

Storing Data

To keep from losing the work you do on a computer, you must save it to some kind of storage medium. Storage devices may include a hard disk drive and various types of removable cartridges and disks. The amount of storage space available on a hard disk or other device is important because this is where you will store the files you create.

Most hard disks are permanently connected to the drive (fixed disks), although there are also removable hard disks. Other removable devices include floppy disks, CDs, DVDs, and magnetic tape cartridges. Each type of removable storage medium works with a corresponding type of drive. Magnetic tape is often used for backup or archiving purposes.

The computer also has internal storage space called *memory* that it uses while it's operating. An adequate amount of memory makes it possible for the computer to easily handle large files while they are open and you are working on them.

Two types of memory are used for internal storage: RAM and ROM. RAM stands for *random-access memory* and is used for temporary storage. ROM stands for *read-only memory*. This is permanent storage for programs that the computer needs to run. It cannot be changed.

Staying Organized

You will be able to work much faster if your files and folders are organized. Follow these tips for managing files.

- Give your files and folders names that describe their contents.
- Avoid storing large numbers of files in a single folder. Organize them into several folders within that folder.
- Save your work as you go. It is recommended that you save every fifteen minutes. Remember, though, that each time you save a file without renaming it, the previous version of your file is overwritten with the newer one.
- Regularly back up your files onto removable storage media, such as floppy disks, tapes, or writable compact discs. (If not preformatted, floppy disks must be formatted before use.) Keep the disks or tapes in a safe place, protected from extremes of temperature and humidity.
- To free up space on the hard disk, delete files that you are sure you no longer need. If you don't need the file now, but may need it later, copy the file to a backup disk or tape. It also helps to delete files that have been stored in the Recycle Bin.

Computer Care and Maintenance

❑ The computer system should be placed on a sturdy, level surface.

❑ Do not block any of the ventilation slots. The computer and monitor should have at least 4" of space around the sides and back for heat to escape.

❑ Keep the computer system away from direct sunlight and heating devices. Avoid dust, high humidity, and extreme heat or cold.

❑ Make sure there is surge protection.

❑ Keep liquids away from the computer and keyboard. Keep disks away from magnets.

❑ Restart the computer every morning, even if you left it on overnight. This clears out all of the memory. Also, if some applications fail to quit completely, this closes them out.

❑ About once a month, run the utilities for defragmenting and diagnosing problems on the hard disk. For Windows 2000, the tools to use are Disk Defragmenter and Disk Cleanup. You can start these utility programs from the Start, Programs, Accessories, and System Tools menus.

❑ Protect the computer against viruses with an antivirus utility program. Two popular programs are Norton Antivirus by Symantec and McAfee VirusScan.

Internet Permission Contract

This school is fortunate to have been equipped with access to the Internet. Our system has been established to help students learn and to help them develop their computer skills. It is our school's intention to use the computer(s) and the Internet responsibly and ethically.

Below is a list of rules and principles we intend to abide by and maintain.

1. Students will have access to the Internet, including e-mail.
2. No personal contact information (such as the student's home and school addresses, telephone numbers, etc.) will be posted or shared on the Internet.
3. Students will report any inappropriate messages to the instructor.
4. Students will not engage in any illegal act, including downloading illegal music, movies, or software and plagiarizing content from the Internet.
5. Students will not attempt to gain unauthorized access or disrupt any computer system by willfully destroying data or spreading viruses.
6. On the Internet, students will not use obscene language, engage in personal or discriminatory attacks, or post false or misleading information about individuals or organizations.
7. Students will not access lewd or obscene material.
8. Students will abide by any additional rules posted by the school.

These rules and principles must be adhered to and any violation of them will be met with zero tolerance.

- Students must abide by their signed contracts.
- The use of the computer is a privilege, not a right.
- If this contract is broken, the student's privilege will be revoked.

I have read the "Internet Permission Contract" above. By signing below, I fully understand and agree with the contents of this contract. If I breach any rules, I understand my Internet and computer privileges will be revoked.

Signed by:

Parent or Guardian: _____ Date:_____

Student: _____ Date:_____

Instructor: _____ Date:_____

Internet Evaluation Sheet

URL of website you are evaluating: http://
_____
_____

Appropriateness: Is the site suitable for viewing? Does it provide the information you need? Is the information at your level and clearly written?
_____
_____
_____

Design and Technical Aspects: Is it attractive and easy to use? Are pages set up in a useful order? Does it download quickly? Do its links work? Are they clearly explained and relevant?
_____
_____
_____

Source: Is the information from a source you can trust? Who wrote the information? Does the site provide biographical information, such as the author's position, the organization the person is associated with, and the address, e-mail address, and telephone number?
_____
_____
_____

Accuracy: If facts or statistics are provided, does the author cite the sources so you can refer to them? How recent is the information? Do the links reveal any bias that the author might have?
_____
_____
_____

Point of View: Is the site's intention to teach something or to persuade you to adopt its point of view? What do you think the writer's point of view, or bias, is? Is the purpose of the site to sell a product?
_____
_____
_____

Using the Internet

The Internet is a worldwide network of computers linked together. Small groups of computers, such as those at a university, may be linked to form a local network. Then that network is linked to other small networks to form a large network. Many large networks make up what we know as the Internet. Using your own computer, you can access any of these networked computers until you find information you are looking for or a person you wish to contact.

E-Mail

Electronic mail (e-mail) is one part of the Internet. It has become an extremely popular method of communication. Using your computer, you can send and receive messages, letters, and documents 24 hours a day. For example, you can write to friends, inquire about products, and send your resume to apply for a job.

As with regular mail, e-mail must include an address. An e-mail address looks something like this:

john_smith@company.com

The World Wide Web

The World Wide Web is another part of the Internet that was created to make accessing information easier. It does this by using hypertext markup language (HTML), a formatting system that allows you to receive information in the form of pictures, text, sound, and even video. Because it is easy to use, the Web is very popular.

Any individual or organization can set up a Web site, a document on the Web, that other people can visit using their computers. A Web site may have several "pages," like the pages in a book, filled with information, pictures, and links to other Web sites. The primary page is called the *home page*. Each page has its own address, called a *uniform resource locator* (URL). By typing that address in your computer, you can go to that Web page.

Search Tools

Newspaper and magazine articles, automobile ratings, travel information, and much, much more can be found on the Internet. Search tools, such as search engines and directories, organize lists of information to make it easier for you to find what you're looking for. For example, Google is a search engine that uses special software to organize and help you find the information you need. A directory, such as Yahoo, is created by people who visit and evaluate Web sites and then organize them into subject-based categories and subcategories. Yahoo also provides a search tool that can be used to find information in the categories.

To use Google, for example, you first type in its URL:

www.google.com

When the home page appears, you can type in key words that describe the information you want. For example, if you are looking for information about engineered lumber, you might type in:

engineered lumber

A list of related links will appear on your screen. If the list you get has links that are not really related to what you're looking for, you can refine your search. Most search tools provide instructions that will save you time and effort. Also, you may get clues for better search terms from words that appear in the links.

Viruses

A virus is a computer program designed to change the way your computer works. Some viruses are relatively harmless. Others may affect system operation, damage other programs, erase files, erase your hard disk, or cause the system to crash.

Most viruses enter your computer through e-mail attachments. If e-mail with an attachment comes from an unknown sender, or

(Continued on next page)

Using the Internet (Continued)

unexpectedly from a known sender, it should not be opened. If you receive an unexpected attachment from someone you know, call to make sure of the mail's source before opening it.

The best protection is an antivirus utility program designed to track down viruses and eliminate them. Two popular antivirus programs are Norton Antivirus by Symantec and McAfee VirusScan. They provide easily accessible anti-virus updates. There is no need to forward e-mail virus warnings, as almost all are hoaxes.

Internet Communities

You can participate in several types of communities on the Internet, such as chat rooms, bulletin boards, newsgroups, and e-mail lists.

- Chat rooms are formed by groups of people who communicate instantly and directly about a topic of common interest.
- Bulletin boards are like chat rooms, except that they are used to post messages rather than to "talk" back and forth. You might post a question about where to find a specific type of building material. Another user might know the answer and respond.
- Newsgroups are similar to bulletin boards, except that you subscribe in order to participate. Newsgroups exist for a huge variety of topics, to which anyone can post thoughts, opinions, or advice.
- Mailing lists are another type of group that you can subscribe to. Messages sent to the group arrive in each person's e-mail box.

Job Searches

The Internet has become very popular with employers and job seekers. Most large companies now have Web sites and provide a page listing their job opportunities. Most large newspapers and some trade magazines reproduce their "Help Wanted" section at their Web sites.

Commercial Web sites have also been created that allow job seekers to find job openings and to post their résumés. Prospective employers review the résumés and contact those people in whom they're interested. Here are two examples:

www.careerweb.com

www.monster.com

You can also send a résumé with an e-mail message, which many companies prefer.

Protecting Yourself

The Internet is a vast source of information and a great way to communicate with friends and relatives. However, just as in real-life, the virtual life of the Internet has some dangers. To help protect yourself, follow these guidelines:

- Don't give personal information, such as your full name, address, etc., to people or companies you don't know.
- Report inappropriate messages or requests for personal information to an instructor or school officials.
- Never give your password to anyone. If someone claiming to represent your Internet service provider asks for your password, don't comply. Contact the service provider yourself and find out whether the request came from them.
- Don't give your credit card number unless you are sure that you are dealing with a reputable company.
- Don't download or open files that are attached to e-mail unless you know the source. The attached file could contain a virus.
- Don't assume that all the information you find on the Internet is true.
- Things change quickly on the Internet. Sites may be redesigned, updated, or disappear. Be aware that what's here today may be gone tomorrow.
- Be aware that some schools may reserve the right to monitor your use of e-mail and Web sites you visit. This is also very common on the job.

Contents

Chapter	1	The Construction Industry	97
Chapter	2	Building Codes and Planning	99
Chapter	3	Reading and Drawing Plans	101
Chapter	4	Estimating and Scheduling	103
Chapter	5	Construction Safety and Health	105
Chapter	6	Hand Tools	107
Chapter	7	Power Saws	109
Chapter	8	Electric Drills	113
Chapter	9	Power Tools for Shaping and Joining	115
Chapter	10	Power Nailers and Staplers	117
Chapter	11	Ladders and Scaffolds	119
Chapter	12	Concrete As a Building Material	121
Chapter	13	Locating the House on the Building Site	123
Chapter	14	Foundation Walls	125
Chapter	15	Concrete Flatwork	127
Chapter	16	Wood As a Building Material	129
Chapter	17	Engineered Lumber	131
Chapter	18	Engineered Panel Products	133
Chapter	19	Framing Methods	135
Chapter	20	Floor Framing	137
Chapter	21	Wall Framing and Sheathing	139
Chapter	22	Basic Roof Framing	141
Chapter	23	Hip, Valley, and Jack Rafters	143
Chapter	24	Roof Assembly and Sheathing	145
Chapter	25	Roof Trusses	147

Contents (Continued)

Chapter 26	Steel Framing Basics	149
Chapter 27	Steel Framing Methods	151
Chapter 28	Windows and Skylights	153
Chapter 29	Residential Doors	155
Chapter 30	Roof Coverings	157
Chapter 31	Roof Edge Details	159
Chapter 32	Siding	161
Chapter 33	Brick-Veneer Siding	163
Chapter 34	Stairways	165
Chapter 35	Molding and Trim	167
Chapter 36	Cabinets and Countertops	169
Chapter 37	Wall Paneling	171
Chapter 38	Mechanicals	173
Chapter 39	Thermal and Acoustical Insulation	175
Chapter 40	Walls and Ceilings	177
Chapter 41	Exterior and Interior Finishes	179
Chapter 42	Wood Flooring	181
Chapter 43	Resilient Flooring and Ceramic Tile	183
Chapter 44	Chimneys and Fireplaces	185
Chapter 45	Decks and Porches	187

Instructional Plan

Focus & Planning

Objectives
- Describe career specialties in construction.
- Describe educational and training programs that can prepare you for a construction career.
- Explain the function of a business plan for entrepreneurs.
- List the primary employability skills.
- Describe the steps in obtaining a job.
- List several responsibilities of employers and employees.

SAFETY FIRST
- ❏ Review the safety rules in the chapter.
- ❏ Make sure all necessary personal protective equipment is available and in good condition.
- ❏ Make sure all required tools and equipment are available and in good working order.

Instruction & Student Practice

Using the *Student Textbook*, have students:
- ❏ Read Section 1.1.
- ❏ Complete Section 1.1 Check Your Knowledge and On the Job.
- ❏ Read Section 1.2.
- ❏ Complete Section 1.2 Check Your Knowledge and On the Job.
- ❏ Read Section 1.3.
- ❏ Complete Section 1.3 Check Your Knowledge and On the Job.
- ❏ Read Section 1.4.
- ❏ Complete Section 1.4 Check Your Knowledge and On the Job.
- ❏ Read Section 1.5.
- ❏ Complete Section 1.5 Check Your Knowledge and On the Job.
- ❏ Read Section 1.6.
- ❏ Complete Section 1.6 Check Your Knowledge and On the Job.

Using *Carpentry Applications*, assign:
- ❏ 1-1: Thinking about a Career
- ❏ 1-2: Creating a Resume

Using *Carpentry Math*, assign:
- ❏ M1-1: Working with Whole Numbers

Using the *Safety Guidebook*, assign:
- ❏ 1-1: Federal Labor Laws
- ❏ 1-4, 1-5, 1-7, 1-8. These relate to safety and health on the job.

Using the *Instructor Resource Guide*, have students:
- ❏ Read the handouts related to student organizations, pp. 41-48.
- ❏ Read Teamwork, pp. 52-54.
- ❏ Read Successful Employee Characteristics and complete the Self Assessment, pp. 50-51.
- ❏ Read Apprenticeship Training, p. 55.
- ❏ Read the handouts related to work-based learning, pp. 56-65.
- ❏ Read the handouts related to All Aspects of Industry, pp. 66-74.
- ❏ Read the handouts related to finding and applying for a job, changes on the job, and entrepreneurship, pp. 75-86.

(Continued on next page)

Review & Student Performance

- ❏ Assign Review Questions from the end of the chapter.
- ❏ Assign the Chapter 1 Test or create a test using the *ExamView® Pro Test Generator* on the *Instructor Productivity CD-ROM*.
- ❏ Create and administer a Unit 1 Pretest using the *ExamView® Pro Test Generator* on the *Instructor Productivity CD-ROM*.
- ❏ Use *Carpentry Applications* job sheets as desired for final performance assessment.

Answers to Textbook Questions

■ Section 1.1 Check Your Knowledge
1. Over 6 million people are employed in construction.
2. They are craft, technical, and professional workers.
3. A trend is a general development or movement in a certain direction.
4. The three factors are family structure, work patterns, and personal preferences.

■ Section 1.1 On the Job
Answers will vary.

■ Section 1.2 Check Your Knowledge
1. Certification is proof of skill.
2. Associate's degree programs are for two years; bachelor's degree programs are for four years and offer general courses.
3. An experienced, skilled worker is the teacher.
4. It is a combination of classroom instruction and work experience.

■ Section 1.2 On the Job
Answers will vary.

■ Section 1.3 Check Your Knowledge
1. *Answers will vary. See Table 1A on p. 31.*
2. It includes a vision, goals, strategies, and a plan of action.
3. Free enterprise is an economic system in which individuals are free to buy, sell, and set prices for goods and services.
4. Zoning is a division of land into areas used for different purposes.

■ Section 1.3 On the Job
Answers will vary.

■ Section 1.4 Check Your Knowledge
1. Reading, writing, mathematics, and science.
2. They include responsibility, flexibility, honesty, cooperation, and commitment.
3. Resources are time, energy, money, materials and equipment, people, and information.
4. Staying up-to-date with new technology makes you valuable to an employer.

■ Section 1.4 On the Job
Answers will vary.

■ Section 1.5 Check Your Knowledge
1. An interview is a meeting between an employer and a job seeker.
2. Sources include classified ads, networking, trade publications, employment agencies, and the Internet.
3. A résumé gives an employer information about your background.
4. Send it the next day.

■ Section 1.5 On the Job
Answers will vary.

■ Section 1.6 Check Your Knowledge
1. Worker's compensation is financial aid from an employer in case of injury on the job.
2. An hourly wage is by the hour; a salary is a set amount per year.
3. Collective bargaining is bargaining done by unions.
4. Ethics are your inner guidelines for telling right from wrong.

■ Section 1.6 On the Job
Answers will vary.

■ Chapter 1 Review Questions
1. They belong to the craft category.
2. Commercial buildings include office and government buildings; residential buildings include homes and apartments.
3. An apprenticeship is the opportunity to learn a trade under the guidance of a skilled worker combined with classroom instruction.
4. A worker is rotated through a series of jobs, which allows the worker to learn a variety of skills.
5. A shareholder is someone who owns shares, or parts of a company.
6. *Answers will vary but should include that it helps the entrepreneur think through the goals and strategies of the business.*
7. The purpose may be to inform or give instructions; to request information, a decision, or an action; to persuade; or to complain or protest.
8. The steps include gathering information, applying for the job, going on an interview, and responding to the offer.
9. He or she can use time responsibly, respect the rules, work safely, and earn his or her pay.
10. Gross pay is the amount earned before deductions; net pay is what remains after deductions.

Instructional Plan

Focus & Planning

Objectives
- Identify the major model building codes used in the United States.
- List the steps in planning to build a house.
- Name three sources of house plans.
- Explain how financing for construction is obtained.

SAFETY FIRST
- ❑ Review the safety rules in the chapter.
- ❑ Make sure all necessary personal protective equipment is available and in good condition.
- ❑ Make sure all required tools and equipment are available and in good working order.

Instruction & Student Practice

Using the *Student Textbook*, have students:
- ❑ Read Section 2.1.
- ❑ Complete Section 2.1 Check Your Knowledge and On the Job.
- ❑ Read Section 2.2.
- ❑ Read Estimating Construction Costs and complete Section 2.2 Estimating on the Job.
- ❑ Complete Section 2.2 Check Your Knowledge and On the Job.

Using *Carpentry Applications*, assign:
- ❑ 2-1: Practicing Design and Problem Solving
- ❑ 2-2: Obtaining a Building Permit

Using *Carpentry Math*, assign:
- ❑ M2-1: Working with Decimals
- ❑ M2-2: Adding, Subtracting, Multiplying, and Dividing Decimals

Review & Student Performance

- ❑ Assign Review Questions from the end of the chapter.
- ❑ Assign the Chapter 2 Test or create a test using the *ExamView® Pro Test Generator* on the *Instructor Productivity CD-ROM*.
- ❑ Use *Carpentry Applications* job sheets as desired for final performance assessment.

(Continued on next page)

Answers to Textbook Questions

■ Section 2.1 Check Your Knowledge

1. A building code is a set of regulations that governs the details of construction. Its purpose is to ensure that buildings are structurally sound and safe from fire and health hazards.
2. A model building code is a set of regulations developed by an independent organization that state and local building departments can adopt or use as a guide to enacting their own codes.
3. A building permit is a formal, printed authorization for the builder to begin construction. You apply for a permit by submitting a full set of working drawings, called plans, to the local building department.
4. A certificate of occupancy (CO) indicates that a house is ready to live in.

■ Section 2.1 On the Job

Answers will vary.

■ Section 2.2 Check Your Knowledge

1. The four legal documents are: the official survey, the deed, the abstract of title, and the contract of sale.
2. A mortgage is a long-term loan that is secured by property. It allows the lender to claim the property if the borrower does not make payments.
3. The percentage of the purchase price of a house that can usually be financed is 80 percent.
4. Actual construction costs can vary because of special features, unusual materials, or custom products.

■ Section 2.2 On the Job

Their annual income is $55,200. The maximum amount they would be able to spend would be $13,800 annually, or $1150 per month.

■ Section 2.2 Estimating on the Job

Answers will vary.

■ Chapter 2 Review Questions

1. The three major model building codes used in the United States include the Uniform Building Code, the National Building Code, and the Standard Building Code.
2. The IRC is a code jointly produced by several of the major model code organizations to cover regional limitations.
3. *Answers will vary.*
4. The builder must contact the building department to schedule an inspection when each part of the job is complete and before it is covered by other work.
5. A general guideline is that not more than 25 percent of gross monthly income should be spent for all housing expenses, including mortgage payments, utilities, and repairs.
6. Six important factors to be considered are cost, location, special conditions, the lot shape and contour, zoning restrictions, and deed restrictions.
7. House plans can be purchased from a company that specializes in designing stock plans, from local builders, or by hiring an architect.
8. A floor plan is a scale drawing showing the size and location of rooms on a given floor.
9. He or she might charge between $12,500 and $25,000.
10. Mortgages usually run from 15 to 30 years.

Instructional Plan

Focus & Planning

Objectives
- Define *scale* and tell how it is used in architectural drawings.
- Name and describe several elements used in architectural drawings.
- Tell the advantages of computer-aided drafting and design.
- Name several of the drawings found in a set of architectural plans and tell their purpose.

SAFETY FIRST
- ❑ Review the safety rules in the chapter.
- ❑ Make sure all necessary personal protective equipment is available and in good condition.
- ❑ Make sure all required tools and equipment are available and in good working order.

Instruction & Student Practice

Using the *Student Textbook*, have students:
- ❑ Read Section 3.1.
- ❑ Complete Section 3.1 Check Your Knowledge and On the Job.
- ❑ Read Section 3.2.
- ❑ Complete Section 3.2 Check Your Knowledge and On the Job.
- ❑ Read Section 3.3.
- ❑ Complete Section 3.3 Check Your Knowledge and On the Job.

Using *Carpentry Applications*, assign:
- ❑ 3-1: Drawing a Site Plan
- ❑ 3-2: Reading Building Plans

Using *Carpentry Math*, assign:
- ❑ M3-1: Reading a Tape or a Ruler
- ❑ M3-2: Identifying the Scales on the Architect's Scale
- ❑ M3-3: Reading an Architect's Scale

Review & Student Performance

- ❑ Assign Review Questions from the end of the chapter.
- ❑ Assign the Chapter 3 Test or create a test using the *ExamView® Pro Test Generator* on the *Instructor Productivity CD-ROM*.
- ❑ Use *Carpentry Applications* job sheets as desired for final performance assessment.

(Continued on next page)

Answers to Textbook Questions

■ Section 3.1 Check Your Knowledge
1. Another term is working drawings or construction drawings.
2. It is based on the unit of ten.
3. A scale drawing is one drawn with the same proportions as the object it represents, but usually much smaller. Scale represents the ratio between the size of the object as drawn and its actual size. An architect can represent any size building on a single piece of paper by drawing it to a certain scale.
4. The size is one forty-eighth or 1:48.

■ Section 3.1 On the Job
The exact length of the line would be 20' 7 $\frac{1}{16}$". However, you could not measure it exactly using the architect's scale. The closest measurement would be approximately 20' 7".

■ Section 3.2 Check Your Knowledge
1. A centerline is composed of long and short dashes, alternately and evenly spaced with a long dash at each end; at intersections the short dashes cross. Centerlines are used to indicate the center of an object.
2. They are lines that mark the end points of a dimension.
3. They usually represent objects and materials.
4. CADD stands for computer-aided drafting and design. These programs are used to create site plans, floor plans, elevation drawings, and even perspective drawings of a structure.

■ Section 3.2 On the Job
Answers will vary.

■ Section 3.3 Check Your Knowledge
1. The four types are plan views, elevations, section views, and detail drawings.
2. Typical sections show features that are repeated many times throughout a structure. They are labeled "TYP" or "Typical."
3. *Any three of the following:* You would expect to find information about the glazing, finish, trim, manufacturer's name, window type, and a letter or number that keys the information to one of the drawings.
4. Materials and finishes for floors, walls, and ceilings are found on a room-finish schedule.

■ Section 3.3 On the Job
There will be three doors, including the garage door.

■ Chapter 3 Review Questions
1. A sketch is a quick and informal drawing.
2. Scale is the ratio between the size of an object as drawn and its actual size. Proportions must be the same. Scale allows large buildings to be drawn on small sheets of paper.
3. A scale of ¼" = 1'-0" is most often used for drawing houses.
4. Hidden lines consist of short dashes evenly spaced and are used to show the hidden features of a part.
5. They are included wherever necessary to provide information not clearly indicated in other ways.
6. Revisions can be made without redrawing; estimating software combined with CADD allows materials lists to be made; symbol libraries provide pre-drawn symbols that save time.
7. An elevation allows you to see the height of objects, which you would not see on a plan view. Also, an elevation shows what exterior materials (such as siding) look like, which is not on a plan view.
8. A wall framing elevation would show the location of studs, plates, sills, and bracing in a wall, because such details cannot be shown completely in a framing plan.
9. The purpose is to show what the completed building will look like.
10. Specifications are a list of written notes that give detailed instructions regarding materials and methods of work, particularly those regarding quality standards.

Instructional Plan

CHAPTER 4: Estimating and Scheduling

Focus & Planning

Objectives
- Name and describe the three basic types of cost estimates.
- Give an example of a direct cost and an indirect cost.
- Draw a simple critical path method (CPM) diagram.

SAFETY FIRST
- ❑ Review the safety rules in the chapter.
- ❑ Make sure all necessary personal protective equipment is available and in good condition.
- ❑ Make sure all required tools and equipment are available and in good working order.

Instruction & Student Practice

Using the *Student Textbook*, have students:
- ❑ Read Section 4.1.
- ❑ Complete Section 4.1 Check Your Knowledge and On the Job.
- ❑ Read Section 4.2.
- ❑ Complete Section 4.2 Check Your Knowledge and On the Job.

Using *Carpentry Applications*, assign:
- ❑ 4-1: Planning the Building Sequence of a House
- ❑ 4-2: Scheduling for Building a House

Using *Carpentry Math*, assign:
- ❑ M4-1: Introduction to Fractions
- ❑ M4-2: Adding and Subtracting Fractions

Review & Student Performance

- ❑ Assign Review Questions from the end of the chapter.
- ❑ Assign the Chapter 4 Test or create a test using the *ExamView® Pro Test Generator* on the *Instructor Productivity CD-ROM*.
- ❑ Create and administer a Unit 1 Posttest using the *ExamView® Pro Test Generator* on the *Instructor Productivity CD-ROM*.
- ❑ Use *Carpentry Applications* job sheets as desired for final performance assessment.

(Continued on next page)

Answers to Textbook Questions

■ Section 4.1 Check Your Knowledge

1. It is important because estimates relate directly to the profitability of the contractor's business. If construction costs are consistently underestimated, a contractor will eventually go out of business. If they are consistently overestimated, the contractor may lose jobs to contractors with lower, more accurate bids.
2. An allowance is a dollar figure representing the cost of products that have not yet been chosen when a detailed estimate is made.
3. A direct cost is specifically related to the construction of a particular house. It includes such costs as those for labor, materials, temporary power hookups, and some types of insurance.
4. The information can be based on the builder's standard construction-order process, or it can be based on a standardized format such as the CSI MasterFormat.

■ Section 4.1 On the Job
Answers will vary.

■ Section 4.2 Check Your Knowledge

1. Materials and activities must be scheduled.
2. In chronological order, the tasks are B, F, G, D, E, A, C.
3. A punch list identifies all the repairs that must be completed before the house is acceptable to the owner.
4. The arrows on the schedule indicate tasks. The nodes represent events.

■ Section 4.2 On the Job
Answers will vary.

■ Chapter 4 Review Questions

1. It is important because estimates relate directly to the profitability of the contractor's business. If construction costs are consistently underestimated, a contractor will eventually go out of business. If they are consistently overestimated, the contractor may lose jobs to contractors with lower, more accurate bids.
2. An estimator must: (1) Be able to read and measure building plans accurately. (2) Have an excellent understanding of the materials and techniques used to build houses. (3) Be precise in assembling and computing numerical data.
3. (1) A pre-design estimate is performed before the exact features of the house are known. (2) A quantity takeoff estimate includes every piece of material required to build the house. (3) For a unit-cost estimate, the estimator divides the house into components and establishes costs for each.
4. It is a dollar figure representing the cost of products that have not yet been chosen when a detailed estimate is made.
5. An indirect cost is one that is not specifically related to a particular house. This would include the cost of office equipment and supplies, as well as office payroll.
6. The organization is the Construction Specifications Institute.
7. *The sources are any three of the following:* Prior bids, material suppliers' bids, pricing guides, and databases.
8. This "call before you dig" precaution helps to prevent the accidental severing of buried utility lines that may cross the property.
9. A bar chart can show an overview of the entire project. It also shows how various tasks overlap each other. However, a bar chart cannot show complicated relationships among the various jobs.
10. What activities precede this activity? What activities cannot start until this activity is complete? What activities can be conducted simultaneously?

Instructional Plan

Chapter 5: Construction Safety and Health

Focus & Planning

Objectives
- Describe the purpose of OSHA.
- Identify practices for keeping a worksite safe.
- Recognize the hazards associated with different types of tools.
- Understand the need for protective clothing and personal protective devices that suit various weather and job-site conditions.

SAFETY FIRST
- ☐ Review the safety rules in the chapter.
- ☐ Make sure all necessary personal protective equipment is available and in good condition.
- ☐ Make sure all required tools and equipment are available and in good working order.

Instruction & Student Practice

Using the *Student Textbook*, have students:
- ☐ Read Section 5.1.
- ☐ Complete Section 5.1 Check Your Knowledge and On the Job.
- ☐ Read Section 5.2.
- ☐ Complete Section 5.2 Check Your Knowledge and On the Job.
- ☐ Read Section 5.3.
- ☐ Complete Section 5.3 Check Your Knowledge and On the Job.
- ☐ Read Estimating Clothing Costs and complete Section 5.3 Estimating on the Job.

Using *Carpentry Applications*, assign:
- ☐ 5-1: Demonstrating Safety Awareness
- ☐ 5-2: Preventing Bodily Injury

Using *Carpentry Math*, assign:
- ☐ M5-1: Multiplying Fractions
- ☐ M5-2: Dividing Fractions

Using the *Safety Guidebook*, assign:
- ☐ 1-2 through 1-6 and 3-1 through 3-10. These relate to safety and health on the job and safety on the worksite.

Using the *Instructor Productivity CD-ROM*, show PowerPoint presentation:
- ☐ 5-1: Demonstrating Safety Awareness
- ☐ 5-2: Preventing Bodily Injury

Using the *Instructor Resource Guide*, have students:
- ☐ Read General Power Tool Safety, p. 33-34.
- ☐ Read Developing Fire Emergency Plan, p. 38.

Review & Student Performance

- ☐ Assign Review Questions from the end of the chapter.
- ☐ Assign the Chapter 5 Test or create a test using the *ExamView® Pro Test Generator* on the *Instructor Productivity CD-ROM*.
- ☐ Create and administer a Unit 2 Pretest using the *ExamView® Pro Test Generator* on the *Instructor Productivity CD-ROM*.
- ☐ Use *Carpentry Applications* job sheets as desired for final performance assessment.

(Continued on next page)

Answers to Textbook Questions

■ Section 5.1 Check Your Knowledge

1. The purpose of the Occupational Safety and Health Act is to assure, so far as possible, safe and healthful working conditions and to preserve our human resources.
2. An employer's basic responsibilities under OSHA are to make sure the job and the workplace are free from recognized hazards that are likely to cause death or serious physical harm and to comply with OSHA regulations.
3. An employee's basic responsibilities under OSHA are to comply with OSHA regulations and to comply with other applicable occupational safety and health standards.
4. Falls are the most common cause of injury on a construction site.

■ Section 5.1 On the Job

Answers will vary.

■ Section 5.2 Check Your Knowledge

1. The purpose of a barricade is to prevent people from entering a dangerous area, such as an excavation.
2. Red means "danger or emergency." Orange means "be on guard." Yellow means "caution." White identifies storage areas. Green identifies first-aid equipment. Blue is for information or caution.
3. It is important not to block exits in case of fire, so people can leave the area quickly.
4. In benching, the soil is excavated to form one or more levels, or "steps." The surfaces between levels are vertical or nearly vertical.

■ Section 5.2 On the Job

Answers will vary.

■ Section 5.3 Estimating on the Job

Answers will vary.

■ Section 5.3 Check Your Knowledge

1. Conductors allow electricity to flow through them readily; insulators do not.
2. The five most basic safety rules are the following: (1) Use the right tool for the job. (2) Examine each tool for damage before use and do not use damaged tools. (3) Operate tools according to the manufacturer's instructions. (4) Keep all tools in good condition with regular maintenance. (5) Properly use the right personal protective equipment.
3. MSDS stands for Material Safety Data Sheet (MSDS). These sheets explain how to handle hazardous products.
4. The five ways are the following: (1) Wear suitable clothing. (2) Install portable fans when working indoors. (3) Drink plenty of water—as much as a quart per hour. (4) Alternate work with rest periods in a cool area. If possible, schedule heavy work during the cooler parts of the day. (5) Get used to the heat through short exposures.

■ Section 5.3 On the Job

Answers will vary.

■ Chapter 5 Review Questions

1. OSHA stands for Occupational Safety and Health Administration. OSHA's purpose is to protect the safety and health of America's workers.
2. A danger sign warns people of immediate hazards. A caution sign warns about potential hazards or unsafe practices. The color red signifies danger. Yellow signifies caution.
3. Weeds, grass, and debris must be cleared from a construction site because they pose a fire hazard. Debris can also cause falls.
4. A GFCI is a ground-fault circuit interrupter. This is a fast-acting circuit breaker that can protect people from electrical shock.
5. Fuel vapors can burn or explode, and the engine produces dangerous exhaust fumes.
6. You should secure the work with clamps or a vise whenever possible. Don't hold it in your hand.
7. Repetitive stress injury is damage to nerves and tissues caused gradually from years of impact or motion.
8. Eye injuries may be caused by sparks, by flying particles, or by chemicals or liquids that splash into a worker's face.
9. A respirator protects you against the harmful effects of breathing fumes or very fine dust. A dust mask can filter out large particles but not fumes or very fine dust.
10. *The worker with heat exhaustion still sweats but experiences any three of the following:* Extreme weakness or fatigue, giddiness, nausea, headache, clammy and moist skin, normal or slightly higher body temperature.

Instructional Plan

Focus & Planning

Objectives
- Identify a wide range of hand tools used in carpentry.
- Tell what different hand tools are used for.

SAFETY FIRST
- ❏ Review the safety rules in the chapter.
- ❏ Make sure all necessary personal protective equipment is available and in good condition.
- ❏ Make sure all required tools and equipment are available and in good working order.

Instruction & Student Practice

Using the *Student Textbook*, have students:
- ❏ Read Section 6.1.
- ❏ Complete Section 6.1 Check Your Knowledge and On the Job.
- ❏ Read Section 6.2.
- ❏ Complete Section 6.2 Check Your Knowledge and On the Job.
- ❏ Read Section 6.3.
- ❏ Complete Section 6.3 Check Your Knowledge and On the Job.
- ❏ Read Section 6.4.
- ❏ Complete Section 6.4 Check Your Knowledge and On the Job.
- ❏ Read Section 6.5.
- ❏ Complete Section 6.5 Check Your Knowledge and On the Job.

Using *Carpentry Applications*, assign:
- ❏ 6-1: Using the Framing Square
- ❏ 6-2: Using a Speed® Square

Using *Carpentry Math*, assign:
- ❏ M6-1: Converting Units of Measure

Using the *Safety Guidebook*, assign:
- ❏ 5-1 through 5-6. These relate to safe use of hand tools.
- ❏ 5-9: Powered Abrasive Wheel Tools

Review & Student Performance

- ❏ Assign Review Questions from the end of the chapter.
- ❏ Assign the Chapter 6 Test or create a test using the *ExamView® Pro Test Generator* on the *Instructor Productivity CD-ROM*.
- ❏ Use *Carpentry Applications* job sheets as desired for final performance assessment.

(Continued on next page)

Answers to Textbook Questions

■ Section 6.1 Check Your Knowledge
1. Layout tape measures are used for laying out foundations and site features, such as walkways and driveways.
2. A chalk line is used.
3. A combination square can be used.
4. A scratch awl would be used.

■ Section 6.1 On the Job
Answers will vary.

■ Section 6.2 Check Your Knowledge
1. A utility drywall saw is used.
2. It cuts with the grain.
3. It is a larger version of a keyhole saw.
4. A coping saw has a deep, U-shaped throat.

■ Section 6.2 On the Job
Answers will vary.

■ Section 6.3 Check Your Knowledge
1. A block plane cuts bevel-side up. A jointer plane cuts bevel-side down.
2. A wood rasp has raised teeth instead of cutting ridges.
3. It cuts roof shingles, tarpaper, batt insulation, wood veneer, and many other materials.
4. They are stored in the handle.

■ Section 6.3 On the Job
Answers will vary.

■ Section 6.4 Check Your Knowledge
1. It is the hand sledge.
2. Claw hammers, ripping bars, and pry bars are used to pull nails.
3. A nail set is used.
4. It attaches ceiling tile, screening, and other soft or thin materials.

■ Section 6.4 On the Job
Answers will vary.

■ Section 6.5 Check Your Knowledge
1. Wrenches should be used.
2. An Allen wrench is used.
3. Needle-nose pliers or lineman's pliers are used.
4. They are also called tin snips or aviation snips.

■ Section 6.5 On the Job
Answers will vary.

■ Chapter 6 Review Questions
1. A layout tape measure.
2. *Any three of the following*:
   - To check adjacent surfaces for the correct angle of 45° or 90°.
   - To make layout lines at 45° and 90° across the face or edge of stock.
   - To measure the depth of a mortise or channel.
   - To roughly level or plumb a surface.
3. Crosscut saws cut *across* the grain of wood. Rip saws cut *with* the grain of wood.
4. They have narrow blades.
5. A dovetail saw has a narrower and thinner blade with finer teeth.
6. It is also called a jack plane.
7. A hand sledge (hand drilling hammer) is the best choice.
8. Models used by framing carpenters have longer handles and checkered faces to reduce glancing blows and flying nails.
9. They are used to drive finishing nails below the surface of wood.
10. *Any two of the following*:
    - Slip-joint pliers are all-purpose adjustable pliers for light-duty gripping; they hold and turn round pieces.
    - Needle-nose pliers can hold and bend thin wire and metal fittings and cut light-gauge electrical wire.
    - Box-joint utility pliers are used to hold and turn large, round parts, and as general gripping tools.
    - Lineman's pliers are used to cut electrical and other wire and for twisting or grasping wire.
    - Vise-grip pliers are used to clamp a workpiece.

Instructional Plan

CHAPTER 7: Power Saws

Focus & Planning

Objectives
- List at least five safety rules that apply to each type of power saw.
- Explain the causes of kickback on circular saws and table saws.
- Make a rip cut with a circular saw, a table saw, and a jigsaw.
- Maintain a circular saw.
- Maintain a table saw.
- Choose a suitable blade for each type of power saw and material.

SAFETY FIRST
- ❑ Review the safety rules in the chapter.
- ❑ Make sure all necessary personal protective equipment is available and in good condition.
- ❑ Make sure all required tools and equipment are available and in good working order.

Instruction & Student Practice

Using the *Student Textbook*, have students:
- ❑ Read Section 7.1.
- ❑ Complete Section 7.1 Check Your Knowledge and On the Job.
- ❑ Read Section 7.2.
- ❑ Complete Section 7.2 Check Your Knowledge and On the Job.
- ❑ Read Section 7.3.
- ❑ Complete Section 7.3 Check Your Knowledge and On the Job.
- ❑ Read Section 7.4.
- ❑ Complete Section 7.4 Check Your Knowledge and On the Job.
- ❑ Read Section 7.5.
- ❑ Complete Section 7.5 Check Your Knowledge and On the Job.
- ❑ Read Section 7.6.
- ❑ Complete Section 7-6 Check Your Knowledge and On the Job.

Using *Carpentry Applications*, assign:
- ❑ 7-1: Using an Electric Drill
- ❑ 7-2: Making Common Saw Cuts

Using *Carpentry Math*, assign:
- ❑ M7-1: Adding and Subtracting Denominate Numbers
- ❑ M7-2: Multiplying and Dividing Denominate Numbers by Whole Numbers

Using the *Safety Guidebook*, assign:
- ❑ 5-7: Electric Tools
- ❑ 5-13 through 5-17. These relate to safe use of various types of power saws.

Review & Student Performance

- ❑ Assign Review Questions from the end of the chapter.
- ❑ Assign the Chapter 7 Test or create a test using the *ExamView® Pro Test Generator* on the *Instructor Productivity CD-ROM*.
- ❑ Use *Carpentry Applications* job sheets as desired for final performance assessment.

(Continued on next page)

Answers to Textbook Questions

■ Section 7.1 Check Your Knowledge
1. If the 2×4 cannot be held safely, clamp it to a work surface.
2. Worm-drive saws require a special lubricant to protect the internal gears.
3. Only about ⅛" of the blade should show below the stock.
4. To make a bevel cut, loosen the wing nut or adjusting lever and tilt the saw to the desired angle. Then retighten the wing nut or lever. Adjust the saw for the correct depth of cut. Make the bevel cut freehand or guide it with a protractor.

■ Section 7.1 On The Job
Answers will vary.

■ Section 7.2 Check Your Knowledge
1. The good side should face up.
2. The miter gauge and the rip fence are used to guide stock.
3. You should: (1) Unplug the saw. (2) Remove the throat plate. (3) Select a wrench to fit the arbor nut. Hold a piece of scrap wood against the blade to keep the arbor from turning. (4) Remove the arbor nut and the collar. (5) Remove the old blade. (6) Slip the new blade onto the arbor and replace the collar and nut. (7) Tighten the nut firmly. (8) Replace the throat plate.
4. Use a roller stand to support long stock. An outfeed table can also be placed to support stock.

■ Section 7.2 On The Job
Answers will vary.

■ Section 7.3 Check Your Knowledge
1. The advantage of a radial-arm saw over other saws is its ability to crosscut unusually wide or thick stock.
2. Adjust the anti-kickback device so that it clears the top of the stock by about ⅛".
3. One hand should hold the stock on the table firmly against the guide fence. The other hand should grasp the motor yoke handle to control the saw's movement.
4. A 10" saw is the most common size.

■ Section 7.3 On The Job
Answers will vary.

■ Section 7.4 Check Your Knowledge
1. A conventional miter saw is the lightest type. The head of the saw pivots up and down and can be swiveled from side to side. A compound-miter model has the same range of motion, but the head of the saw can also be tilted to one side or the other. This allows it to make compound-miter cuts in one pass. A sliding compound-miter saw is similar to compound-miter models, except that it has the capacity to cut wider stock. The head of the saw slides back and forth on one or two metal rails, as well as pivoting up and down.
2. A compound-miter cut combines a miter and a bevel. It is often used when cutting crown moldings, handrails, and other trims that require complex fitting.
3. Using the blade brake to stop the blade quickly after a cut is made reduces hand injuries caused by a freely spinning blade.
4. With conventional miter and compound-miter saws, the cut is made by pivoting the blade *down* into the wood. With a sliding compound-miter saw, the blade is pushed *forward* through the wood.

■ Section 7.4 On The Job
Answers will vary.

■ Section 7.5 Check Your Knowledge
1. The jigsaw is sometimes called a saber saw or a bayonet saw. It is the best power saw for making curved or irregular cuts where precise control is necessary.
2. You should move the saw blade slowly into the wood; don't force it. Use only enough pressure to keep the saw cutting at all times.
3. An electric drill fitted with a ⅜" spade bit is required.
4. When finished using a reciprocating saw, turn off the power and allow the saw to come to a stop *before* pulling the blade from the cut and setting the saw down.

■ Section 7.5 On the Job
Answers will vary.

■ Section 7.6 Check Your Knowledge
1. The five basic parts are the arbor, the body, the teeth, the shoulder, and the gullet.
2. The gullet helps remove sawdust from the cut.
3. These blades cut a wider variety of materials and need sharpening less often. However, they must be taken to a professional for sharpening. Carbide is brittle and must be stored carefully.
4. A ¼" universal shank is the most common type.

■ Section 7.6 On the Job
Answers will vary.

■ Chapter 7 Review Questions
1. To ensure safety before crosscutting with a radial-arm saw, adjust the anti-kickback device to clear the top of the work by ⅛".
2. Circular saws are normally used for crosscutting softwood lumber and panel products.
3. Kickback occurs when a spinning blade is momentarily stopped while the saw is under full power. The saw may be twisted to the

Answers to Textbook Questions

side during the cut or pulled backwards. The material on one or both sides of the cut may bend, pinching the saw blade. The saw may encounter a large knot, which suddenly slows the blade.

4. To maintain a circular saw
   - Inspect saw guards frequently to ensure that they are working correctly. Never use a saw with a damaged or missing guard.
   - Check that the baseplate is straight. Sight along the bottom of the plate at the beginning of each day and if the saw has been dropped. If a baseplate cannot be bent back into alignment, replace it.
   - Worm-drive saws require a special lubricant to protect the internal gearing. Drain the old lubricant periodically and replace it with new lubricant as recommended by the manufacturer.
   - Inspect the power cord. A saw's power cord can be damaged by the blade or other abuse. If the damage extends through the cord's outer casing, the cord should be replaced.
5. Miter cuts can be made freehand, or they can be guided by a saw protractor.
6. To maintain a table saw
   - Keep the table clean and smooth, remove patches of rust with metal polish.
   - Scrape off paint drips and glue to prevent them from interfering with the movement of stock.
   - Then unplug the saw, gently brush or blow accumulated sawdust from the blade raising and tilting mechanism beneath the table.
   - Lubricate the mechanism as recommended by the manufacturer.
   - Some saws are driven by a V-belt that links the motor to the arbor. Replace the belt if it becomes worn or damaged.
7. Thin-kerf blades require less power during cutting than standard-kerf blades. This makes them particularly suitable for cordless circular saws.
8. A push stick should be used.
9. A circular saw has a circular blade that spins. A jigsaw has a straight blade that moves up and down.
10. Half the teeth are ground in one direction and half in the opposite direction. The two types of teeth are arranged in an alternating pattern.

Instructional Plan

Chapter 8: Electric Drills

Focus & Planning

Objectives
- List five safety tips for using electric drills.
- Identify the two basic types of electric drills.
- Determine the power rating of an electric drill.
- Tell the uses for different types of drill bits.
- Drill holes in wood and metal.

SAFETY FIRST
- ❑ Review the safety rules in the chapter.
- ❑ Make sure all necessary personal protective equipment is available and in good condition.
- ❑ Make sure all required tools and equipment are available and in good working order.

Instruction & Student Practice

Using the *Student Textbook*, have students:
- ❑ Read Section 8.1.
- ❑ Complete Section 8.1 Check Your Knowledge and On the Job.
- ❑ Read Section 8.2.
- ❑ Complete Section 8.2 Check Your Knowledge and On the Job.

Using *Carpentry Applications*, assign:
- ❑ 8-1: Using an Electric Drill
- ❑ 8-2: Building Sawhorses

Using *Carpentry Math*, assign:
- ❑ M8-1: Calculating the Midpoint

Using the *Safety Guidebook*, assign:
- ❑ 5-7: Electric Tools
- ❑ 5-9: Powered Abrasive Wheel Tools

Review & Student Performance

- ❑ Assign Review Questions from the end of the chapter.
- ❑ Assign the Chapter 8 Test or create a test using the *ExamView® Pro Test Generator* on the *Instructor Productivity CD-ROM*.
- ❑ Use *Carpentry Applications* job sheets as desired for final performance assessment.

(Continued on next page)

Answers to Textbook Questions

■ Section 8.1 Check Your Knowledge

1. You should check the specification plate on the tool for its amperage. Amperage is an approximate measure of how powerful the drill is.
2. They both have wide flutes to remove debris from the hole.
3. This bit is excellent for boring smooth holes with flat bottoms in wood. Forstner bits can bore through end grain with ease.
4. This shallow funnel shape in the wood allows the top of a wood screw to be flush with the wood surface.

■ Section 8.1 On The Job

Answers will vary.

■ Section 8.2 Check Your Knowledge

1. The most important factor is choosing a suitable bit for the material and the job to be done.
2. You should check that the shank is centered between the jaws.
3. It is important to apply the proper pressure because too little pressure will dull the bit; too much pressure may break it.
4. You should check the bit by spinning it in the drill briefly before drilling. The tip of a bent bit will wobble noticeably.

■ Section 8.2 On the Job

Answers will vary.

■ Chapter 8 Review Questions

1. You should clamp small pieces of wood or other material to prevent them from spinning.
2. The two basic types of electric drills are corded and cordless.
3. Drills with a plastic housing insulate the operator against electrical shock.
4. The amperage range for corded drills is 3 – 8 amps.
5. The voltage of the battery indicates the length of time a cordless drill can run between chargings.
6. It guides the bit and prevents it from wandering as the hole is started.
7. A Forstner bit would be suitable.
8. In one operation it will drill the pilot hole, the shank hole, and the countersink.
9. You should clamp a piece of scrap wood behind the workpiece.
10. You should use a cutting lubricant such as lightweight oil. When drilling without lubricant, reduce the bit's rpm to prevent overheating.

Instructional Plan

―― Focus & Planning ――

Objectives
- List at least five general safety rules for the use of shaping and joining tools.
- Identify common uses for routers, sanders, planers, and jointers.
- Explain how each tool is classified in terms of its size.

- ❏ Review the safety rules in the chapter.
- ❏ Make sure all necessary personal protective equipment is available and in good condition.
- ❏ Make sure all required tools and equipment are available and in good working order.

―― Instruction & Student Practice ――

Using the *Student Textbook*, have students:
- ❏ Read Section 9.1.
- ❏ Complete Section 9.1 Check Your Knowledge and On the Job.
- ❏ Read Section 9.2.
- ❏ Complete Section 9.2 Check Your Knowledge and On the Job.
- ❏ Read Section 9.3.
- ❏ Complete Section 9.3 Check Your Knowledge and On the Job.
- ❏ Read Section 9.4.
- ❏ Complete Section 9.4 Check Your Knowledge and On the Job.

Using *Carpentry Applications*, assign:
- ❏ 9-1: Using a Router
- ❏ 9-2: Using a Plate Joiner

Using *Carpentry Math*, assign:
- ❏ M9.1: Converting Decimals
- ❏ M9.2: Converting Fractions

Using the *Safety Guidebook*, assign:
- ❏ 1-5: Safe Work Habits
- ❏ 5-7, 5-19 through 5-23. These relate to safe use of power tools, specifically saws.

―― Review & Student Performance ――

- ❏ Assign Review Questions from the end of the chapter.
- ❏ Assign the Chapter 9 Test or create a test using the *ExamView® Pro Test Generator* on the *Instructor Productivity CD-ROM*.
- ❏ Use *Carpentry Applications* job sheets as desired for final performance assessment.

(Continued on next page)

Answers to Textbook Questions

■ Section 9.1 Check Your Knowledge

1. The motor and base of a fixed-base router always remain stationary during a cut. On a plunge router, the motor is mounted on vertical metal posts and can slide up and down, allowing the spinning bit to be "plunged" into the workpiece.
2. A flush-trimming bit is used to trim plastic laminate to size.
3. The accessory is an edge guide.
4. You should hold the router firmly to resist the starting torque of the motor, which tends to twist the tool.

■ Section 9.1 On the Job
Answers will vary.

■ Section 9.2 Check Your Knowledge

1. The machine is classified by the width and length of its belt.
2. You should move the sander either forward and back or from side to side. Never hold it in one place or it will gouge the wood.
3. (1) The sanding motion is both circular and from side-to-side. (2) The sanding pad is round, not square, and has holes in it for dust extraction.
4. You should sand in the direction of the wood grain, not across it.

■ Section 9.2 On the Job
Four pad-sized pieces can be cut from one sheet.

■ Section 9.3 Check Your Knowledge

1. The size of a jointer is indicated by the maximum width of its cut.
2. It is mounted above the bed of the machine and above the wood.
3. On a jointer, it is below the bed, and below the wood.
4. Portable electric planes are used for trimming and squaring edges and surfacing.

■ Section 9.3 On the Job
Answers will vary.

■ Section 9.4 Check Your Knowledge

1. *Applications for plate joinery in residential construction are any two of the following:* Strengthening joints, assembling molding, joining shelves to site-built cabinetry, butt-joining custom wood flooring that is not end-matched, and joining synthetic countertop materials.
2. Biscuits are often cut from beech.
3. This helps the joint resist shear forces.
4. The three standard sizes are #0, #10, and #20.

■ Section 9.4 On the Job
Answers will vary.

■ Chapter 9 Review Questions

1. When using a belt sander, it is safe to let go of the handles when the belt stops moving.
2. Routers are described by the size of the collet. A portable belt sander is categorized by the length and width of its belt. An orbital sander is classified by the size of its pad. A jointer is categorized by the maximum width of its cut. A planer is classified by the size of the widest board it can surface.
3. It is used to cut joints, shape edges, trim plastic laminate, and cut openings in panels.
4. Belt sanders are used to quickly remove a large amount of material. Orbital sanders are used to smooth a surface prior to painting or finishing. Random-orbit sanders are used for both.
5. Many sanding belts are constructed with a lap seam. If they are installed with the seam in the wrong position, it can be ripped open as the sander is used. This is not a problem with "seamless" belts, which are constructed differently.
6. An edge is said to be jointed when it is at right angles to the face of the board along its entire length.
7. Jointing removes saw marks and ensures a square edge.
8. A planer is used to reduce the thickness of a board, smooth its surface, and make one face parallel to another.
9. As the cutter head revolves, the knives make many small cuts on the surface of the board.
10. They cut slots for biscuits used to strengthen a joint.

Instructional Plan

Power Nailers and...

—— Focus & Planning ——

Objectives

- List five safety rules for using nailers and staplers.
- Name the parts of a pneumatic fastening system.
- Tell the main differences between a pneumatic nailer or stapler and a cordless model.
- Identify the two types of fastener magazines.
- Identify fasteners used with power nailers and staplers.

SAFETY FIRST

- ❏ Review the safety rules in the chapter.
- ❏ Make sure all necessary personal protective equipment is available and in good condition.
- ❏ Make sure all required tools and equipment are available and in good working order.

—— Instruction & Student Practice ——

Using the *Student Textbook*, have students:

- ❏ Read Section 10.1.
- ❏ Complete Section 10.1 Check Your Knowledge and On the Job.
- ❏ Read Section 10.2.
- ❏ Complete Section 10.2 Check Your Knowledge and On the Job.

Using *Carpentry Applications*, assign:

- ❏ 10-1: Maintaining a Nailer
- ❏ 10-2: Using a Nailer Safely

Using *Carpentry Math*, assign:

- ❏ M10-1: Converting Percentages
- ❏ M10-2: Simplifying Fractions

Using the *Safety Guidebook*, assign:

- ❏ 5-8: Pneumatic Tools
- ❏ 5-25: Power Nailers and Staplers

—— Review & Student Performance ——

- ❏ Assign Review Questions from the end of the chapter.
- ❏ Assign the Chapter 10 Test or create a test using the *ExamView® Pro Test Generator* on the *Instructor Productivity CD-ROM*.
- ❏ Use *Carpentry Applications* job sheets as desired for final performance assessment.

(Continued on next page)

Answers to Textbook Questions

■ Section 10.1 Check Your Knowledge

1. The trigger must be pulled and the nose of the tool must be pressed against the workpiece before the tool will fire. This helps to prevent the tool from being fired accidentally.
2. First determine the type and size of fastener needed. Then find a tool that will drive that fastener. Finally, find an air compressor to fit the chosen tool.
3. The major parts of an air compressor include a pump, a motor, an air-storage tank, a tank-pressure gauge, a line-pressure gauge, and a regulator.
4. Air moving through a hose is slowed by friction. The longer the hose, the harder the air compressor must work to overcome the friction.

■ Section 10.1 On the Job

Answers will vary.

■ Section 10.2 Check Your Knowledge

1. Collated fasteners are fasteners arranged into strips or rolls and joined by plastic or paper strips or fine wire.
2. It may not hold as well.
3. Staples are typically categorized by length, width of crown, wire size, and type of point.
4. These tools can hold many more nails.

■ Section 10.2 On the Job

Answers will vary.

■ Chapter 10 Review Questions

1. A power nailer should be pointed at the ground when connecting to a pressurized air hose because the sudden entrance of pressurized air into the tool can cause it to fire.
2. The three basic parts are the tool, the hose, and the air compressor.
3. If the pressure is too low, the fastener may not be driven completely. If the pressure is too high, the fastener may be driven too deep. Excess pressure is also hard on the tool.
4. When the piston is forced downward by the compressed air, the driver blade strikes a fastener and pushes it into the wood at high speed.
5. A pneumatic nailer or stapler uses compressed air to drive fasteners. It must be connected to an air compressor by means of a high-pressure air hose. A cordless nailer or stapler operates quite differently. Instead of compressed air, the driving force is supplied by a small internal combustion engine fueled with compressed liquefied gas. The fuel is ignited by a spark from a rechargeable battery in the tool's handle.
6. If recommended by the manufacturer, gaskets should be lubricated on a regular basis. One way to do this is to put a few drops of tool oil into the air intake of the nailer, just before connecting the hose. Another method is to attach a line lubricator to the air compressor. The magazine should be cleaned of dirt and sawdust as needed. Spray the magazine with a lightweight lubricant recommended by the manufacturer, then wipe it clean.
7. Most nailers and staplers operate on pressures of 60 to 120 psi.
8. The basic types of nail shanks used with nailers are: (1) Smooth-shank nails provide good holding power in a variety of woods. (2) Screw-shank nails have more holding power than smooth-shank nails. They have a spiral shape that is particularly useful for nailing hardwoods. (3) Ring-shank nails have a series of ridges or "rings" running from the point nearly to the head. These nails are best for applications that require extra holding power, such as nailing wood that has a high moisture content. Ring-shank nails are sometimes used to nail subfloors because they can reduce the occurrence of squeaks.
9. Nailers may be loaded with nails that are collated. Collated fasteners are arranged into strips or rolls. Staplers may be loaded with staples made of various metals, including steel, galvanized steel, stainless steel, aluminum, and bronze.
10. The two types are strip-loaded and coil-loaded.

Instructional Plan

Scaffolds

Focus & Planning

Objectives
- List five safety tips for working with ladders.
- Tell how to set up a straight ladder and a stepladder safely.
- Compare manufactured metal scaffolding to wood scaffolding.
- Describe brackets, a pump jack, and a lifeline.

SAFETY FIRST
- ❏ Review the safety rules in the chapter.
- ❏ Make sure all necessary personal protective equipment is available and in good condition.
- ❏ Make sure all required tools and equipment are available and in good working order.

Instruction & Student Practice

Using the *Student Textbook*, have students:
- ❏ Read Section 11.1.
- ❏ Complete Section 11.1 Check Your Knowledge and On the Job.
- ❏ Read Section 11.2.
- ❏ Complete Section 11.2 Check Your Knowledge and On the Job.

Using *Carpentry Applications*, assign:
- ❏ 11-1: Using a Ladder Safely
- ❏ 11-2: Building a Ladder

Using *Carpentry Math*, assign:
- ❏ M11-1: Solving Percentage Application Problems
- ❏ M11-2: Calculating Discount Percentages

Using the *Safety Guidebook*, assign:
- ❏ 4-1: Ladders
- ❏ 4-2: Scaffolds

Review & Student Performance

- ❏ Assign Review Questions from the end of the chapter.
- ❏ Assign the Chapter 11 Test or create a test using the *ExamView® Pro Test Generator* on the *Instructor Productivity CD-ROM*.
- ❏ Create and administer a Unit 2 Posttest using the *ExamView® Pro Test Generator* on the *Instructor Productivity CD-ROM*.
- ❏ Use *Carpentry Applications* job sheets as desired for final performance assessment.

(Continued on next page)

Answers to Textbook Questions

■ Section 11.1 Check Your Knowledge

1. The two basic types of ladders are folding ladders and straight ladders. With folding ladders, the legs are spread apart for stability and they are self-supporting. With straight ladders, they are not self-supporting, but must be leaned against a wall or some other object.
2. You should not try to repair a damaged ladder. Replace it or have it professionally repaired.
3. The two types of planks are laminated wood planks made specifically for scaffolding and aluminum planks.
4. Metal scaffolding is engineered and tested to withstand a specific load. It can be rented as needed, takes up less space than wood scaffolding, and is more weather resistant. It is also easier to assemble and disassemble.

■ Section 11.1 On the Job
Answers will vary.

■ Section 11.2 Check Your Knowledge

1. *Answers may vary.* These are often preferred when clearances are restricted around the structure or when working atop the roof.
2. *Answers may vary.* A pump jack has foot pedals that allow the worker to pump each jack upward along the posts. Brackets are more difficult to reposition.
3. No, only trestles should be used to support scaffolding.
4. A lifeline is a rope intended to prevent a worker from falling more than 6'

■ Section 11.2 On the Job
Answers will vary.

■ Chapter 11 Review Questions

1. If you must place a ladder in front of a door or other opening, secure the ladder or use a barricade to keep traffic away.
2. To set up a straight ladder, place the lower end against a base so it cannot slide. Grasp a rung at the top end and walk forward under the ladder, moving your hands to grasp other rungs as you proceed. When the ladder is erect, lean it forward into the desired position. Check the angle, height, and stability at top and bottom. When using a ladder to access a roof, make sure the ladder extends above the edge of the roof by at least 3'. Before using a stepladder, always be certain that the feet are firmly supported and that the spreader is locked into position.
3. The foot of the ladder should be positioned away from the support a distance equal to one-fourth the ladder's working length. The angle created is approximately 75°.
4. If you stand on the top step, your weight can easily unbalance it.
5. A scaffold is a raised platform used for working at a height.
6. A duplex-head nail has a double head. It can be driven in tightly and still be easily pulled out.
7. Metal scaffolding is engineered and tested to withstand a specified load. It can be rented as needed, takes up less space than wood scaffolding, and is more weather resistant. It is also easier to assemble and disassemble. The end frames may be assembled in a staggered position, making it possible to work from a stairway.
8. The three main types of brackets are wall, corner, and roofing brackets.
9. A pump jack is a metal device that slides up and down along wood or metal posts. The jacks support horizontal planks that support the worker. They have foot pedals that allow the worker to pump them upward. To lower the jack, the worker turns a hand crank.
10. For a lifeline, use a manila rope at least ¾" in diameter and able to support a minimum deadweight of 5,400 lbs.

Instructional Plan

Focus & Planning

Objectives
- List the characteristics of concrete that make it a useful construction material.
- List the basic ingredients of concrete.
- List the five basic types of cement.
- Mix a small batch of concrete from a pre-mix.

SAFETY FIRST
- ❑ Review the safety rules in the chapter.
- ❑ Make sure all necessary personal protective equipment is available and in good condition.
- ❑ Make sure all required tools and equipment are available and in good working order.

Instruction & Student Practice

Using the *Student Textbook*, have students:
- ❑ Read Section 12.1.
- ❑ Complete Section 12.1 Check Your Knowledge and On the Job.
- ❑ Read Section 12.2.
- ❑ Complete Section 12.2 Check Your Knowledge and On the Job.
- ❑ Read Section 12.3.
- ❑ Complete Section 12.3 Check Your Knowledge and On the Job.

Using *Carpentry Applications*, assign:
- ❑ 12-1: Making a Concrete Slump Test
- ❑ 12-2: Comparing Concrete Costs

Using *Carpentry Math*, assign:
- ❑ M12-1: Measuring Perimeter
- ❑ M12-2: Measuring Area
- ❑ M12-3: Measuring Volume

Using the *Safety Guidebook*, assign:
- ❑ 2-2: Personal protective Equipment
- ❑ 6-3: Hazardous Materials

Review & Student Performance

- ❑ Assign Review Questions from the end of the chapter.
- ❑ Assign the Chapter 12 Test or create a test using the *ExamView® Pro Test Generator* on the *Instructor Productivity CD-ROM*.
- ❑ Create and administer a Unit 3 Pretest using the *ExamView® Pro Test Generator* on the *Instructor Productivity CD-ROM*.
- ❑ Use *Carpentry Applications* job sheets as desired for final performance assessment.

(Continued on next page)

Answers to Textbook Questions

■ Section 12.1 Check Your Knowledge
1. It is called mortar or grout.
2. It is small lump formed in the kiln after the basic ingredients of cement have been heated to 2,700°F [1482°C].
3. This cement gains strength faster than other types. This is helpful where the forms must be removed quickly or when the concrete must be put into service quickly.
4. An admixture is an ingredient added to the concrete that changes the physical or chemical characteristics of the concrete.

■ Section 12.1 On the Job
Answers will vary.

■ Section 12.2 Check Your Knowledge
1. Strength, durability, watertightness, and wear resistance are controlled by the amount of water in proportion to the amount of cement.
2. When mixed with water, a 60-lb. sack yields 1 cubic foot of concrete.
3. A slump test tests the consistency of fresh concrete.
4. In hot weather, protect concrete from rapid drying. Rapid drying lowers its strength and may damage the exposed surfaces of sidewalks and drives.

■ Section 12.2 On the Job
1. You would need 600 lbs. of fine aggregate and 900 lbs. of coarse aggregate.
2. You would need 200 lbs. of cement and 600 lbs. of coarse aggregate.

■ Section 12.3 Check Your Knowledge
1. When steel is embedded in concrete, the resulting material, called reinforced concrete, has both compression and tensile strength.
2. The deformed surface helps the concrete grip the steel.
3. Reinforcing steel can be purchased in the form of reinforcing bars, known as rebar, or welded-wire fabric. Rebar is used most often in footings and walls, while welded-wire fabric is used mostly in slabs.
4. The most common welded-wire fabric on a residential job site has wires spaced 6" apart in two directions. It is called 6×6 welded-wire reinforcement.

■ Section 12.3 On the Job
Answers will vary.

■ Chapter 12 Review Questions
1. *Concrete has any four of the following characteristics:*
   - It has tremendous compressive strength.
   - It is resistant to chemicals.
   - It will not rot or be damaged by insects.
   - It hardens under a variety of conditions, including under water.
   - When properly cured, it withstands extreme heat and cold.
   - It can be formed into almost any shape.
   - It is widely available and fairly inexpensive.
2. Concrete is made by mixing cement, fine aggregate (usually sand), coarse aggregate (usually gravel or crushed stone), and water in the proper proportions.
3. Type I (standard) is the type used for general construction purposes. It is economical and has a long setting time.
4. Hydraulic cement is used to plug holes and cracks in foundations.
5. All aggregates should be clean and free of dirt, clay, or vegetable matter, which would reduce the strength of the concrete.
6. Hydration is a chemical reaction between the water and the cement that causes the concrete to harden.
7. The strength of concrete is based on the ingredients, the curing methods, and the curing time.
8. Because fresh concrete is highly alkaline, it can irritate the skin and eyes. Wear gloves, rubber boots, and eye protection when mixing and placing concrete.
9. Ready-mix concrete may be ordered by the number of bags of cement per cubic yard of concrete or by its desired compressive strength.
10. Fibers help to reduce shrinkage cracking that sometimes occurs as concrete cures. They also increase concrete's resistance to impact and abrasion.

Instructional Plan

Focus & Planning

Objectives
- Establish a simple building layout, working from an existing reference line.
- Identify the basic types of surveying instruments and list their limitations.
- Use a builder's level to lay out a right angle.
- Describe how to set up batter boards.
- Measure a difference in elevation between two points, using a level or transit.
- Estimate the volume of soil excavated for a house foundation.

SAFETY FIRST
- ☐ Review the safety rules in the chapter.
- ☐ Make sure all necessary personal protective equipment is available and in good condition.
- ☐ Make sure all required tools and equipment are available and in good working order.

Instruction & Student Practice

Using the *Student Textbook*, have students:
- ☐ Read Section 13.1.
- ☐ Complete Section 13.1 Check Your Knowledge and On the Job.
- ☐ Read Section 13.2.
- ☐ Read Estimating Excavation Volume and complete Section 13.2 Estimating on the Job.
- ☐ Complete Section 13.2 Check Your Knowledge and On the Job.

Using *Carpentry Applications*, assign:
- ☐ 13-1: Calculating Diagonals
- ☐ 13-2: Using an Optical Level or Transit

Using *Carpentry Math*, assign:
- ☐ M13-1: Reading a Leveling Rod

Using the *Safety Guidebook*, assign:
- ☐ 1-6, 3-2, 3-8, 3-9, 4-4. These relate to hazardous areas and electrical safety.
- ☐ 5-24: Lasers

Using the *Instructor Productivity CD-ROM*, show PowerPoint presentation:
- ☐ 13-1: Calculating Diagonals

Review & Student Performance

- ☐ Assign Review Questions from the end of the chapter.
- ☐ Assign the Chapter 13 Test or create a test using the *ExamView® Pro Test Generator* on the *Instructor Productivity CD-ROM*.
- ☐ Use *Carpentry Applications* job sheets as desired for final performance assessment.

(Continued on next page)

Answers to Textbook Questions

■ Section 13.1 Check Your Knowledge

1. A plot plan is the part of the house plans that shows the location of the building on the lot, along with related land elevations.
2. An advantage is that the user does not need to look through the instrument, so a single person can set up the laser, move to another portion of the site, and determine level by holding an electronic detector against a leveling rod. The disadvantage of a laser survey instrument is that the light can be difficult to see in bright daylight.
3. The telescope of a level is fixed in a horizontal plane. It can be used only for measuring horizontal angles because it cannot be tilted up and down. The telescope of a transit can be moved up and down as well as from side to side.
4. A plumb bob suspended from the hook is used to center the level or transit directly over the station mark.

■ Section 13.2 Estimating on the Job

456 cubic yards

■ Section 13.2 Check Your Knowledge

1. Sometimes the bench mark is used as a reference point for establishing the grade line. At other times this line may be located in relation to the level of an existing street or curb.
2. A batter board is a board fastened horizontally to stakes placed to the outside of where the corners of the building will be located. These boards and string tied between them mark the outline of the building.
3. Set up a level or transit. Place a leveling rod upright on the first point to be checked. Then sight through the telescope at the leveling rod. Take a reading at that point and then move the rod to the second point. Raise or lower the rod until the reading is the same as for the first point. The bottom of the rod is then at the same elevation as the original point.
4. Excavations must be wide enough to provide space to work. For example, there must be enough room to install and remove concrete forms, to lay up block, to waterproof the exterior surfaces of the walls, and to install foundation drainage.

■ Section 13.2 On the Job

1. 176 cubic yards
2. 25 cubic yards (a total of 200 cubic yards)

■ Chapter 13 Review Questions

1. A builder's level, sometimes called a dumpy level, is the least expensive surveying instrument. An automatic level automatically compensates for variations in set up, and is more accurate than a builder's level. A laser level projects an intense beam of light in a horizontal plane. Some models continually rotate atop a tripod. Others can be automatically rotated using a remote control.
2. It can measure vertical angles. It can be used to determine if a post or wall is plumb. It can establish points along a line.
3. A bench mark is a reference point from which measurements can be made. A station mark is the point over which a surveying instrument has been set up.
4. Set up the instrument directly over the point. Sight a reference point along a base line and set the 360° scale at zero. Turn the telescope until the scale indicates that an arc of 90° has been completed. Use the leveling rod to establish this point along the second line. A right angle is formed where the two lines intersect.
5. Find the setback distance.
6. A plumb bob is used.
7. It is usually best to start by laying out a large rectangle that will encompass all or most of the building. Having established this accurately, the remaining portion of the layout will consist of small rectangles, each of which can be laid out and proved separately.
8. The transit or level is placed at an intermediate point.
9. The first step is to locate the corners of the building precisely and mark them with stakes. Drive nails into the tops of the stakes that indicate the outside line of the foundation walls.
10. 336 cubic yards, rounded off

Instructional Plan

Focus & Planning

Objectives
- Explain the purpose of footings.
- Tell how solid foundation walls are formed and poured.
- Describe the process of laying a concrete block foundation.

SAFETY FIRST
- ❑ Review the safety rules in the chapter.
- ❑ Make sure all necessary personal protective equipment is available and in good condition.
- ❑ Make sure all required tools and equipment are available and in good working order.

Instruction & Student Practice

Using the *Student Textbook*, have students:
- ❑ Read Section 14.1.
- ❑ Read Estimating Concrete and Labor for Footing and complete Section 14.1 Estimating on the Job.
- ❑ Complete Section 14.1 Check Your Knowledge and On the Job.
- ❑ Read Section 14.2.
- ❑ Read Estimating Foundation Walls and complete Section 14.2 Estimating on the Job.
- ❑ Complete Section 14.2 Check Your Knowledge and On the Job.
- ❑ Read Section 14.3.
- ❑ Read Estimating Block Walls and complete Section 14.3 Estimating on the Job.
- ❑ Complete Section 14.3 Check Your Knowledge and On the Job.

Using *Carpentry Applications*, assign:
- ❑ 14-1: Estimating Concrete Blocks
- ❑ 14-2: Framing Footings

Using *Carpentry Math*, assign:
- ❑ M14-1: Estimating a Concrete Footing
- ❑ M14-2: Estimating Block

Using the *Safety Guidebook*, assign:
- ❑ 2-2: Personal Protective Equipment
- ❑ 3-8: Excavations
- ❑ 6-3: Hazardous Materials

Review & Student Performance

- ❑ Assign Review Questions from the end of the chapter.
- ❑ Assign the Chapter 14 Test or create a test using the *ExamView® Pro Test Generator* on the *Instructor Productivity CD-ROM*.
- ❑ Use *Carpentry Applications* job sheets as desired for final performance assessment.

(Continued on next page)

Carpentry & Building Construction Instructor Resource Guide
Copyright © Glencoe/McGraw-Hill

Answers to Textbook Questions

■ Section 14.1 Estimating on the Job
6.2 cubic yards of concrete, 12.2 hours for excavation, and 12.4 hours for placement

■ Section 14.1 Check Your Knowledge
1. It provides a larger bearing surface against the soil.
2. Reinforcing often consists of two lengths of ½" diameter (#4) rebar. The rebar must be positioned at least 3" above the bottom of the footing.
3. Stepped footings are often used where the lot slopes dramatically.
4. A footing drain directs subsurface water away from the foundation. This helps to prevent damp basement walls and wet floors.

■ Section 14.1 On the Job
Answers will vary.

■ Section 14.2 Estimating on the Job
About 88 ½ hours to install forms; 33.6 hours to remove forms; 34 ¾ cubic yards of concrete; 113 hours to place concrete

■ Section 14.2 Check Your Knowledge
1. A wale is a horizontal piece of lumber used to stiffen concrete forms.
2. Insulating concrete forms (ICFs) dramatically increase the insulation value of the foundation walls, a considerable advantage when living space will be located below grade.
3. Adding extra water to the mix weakens the finished walls and encourages cracking.
4. (1) Soil beneath the house must be covered with a material to block moisture. (2) The crawl space usually must be ventilated. Check local codes. (3) The floor framing should be insulated to reduce heat loss.

■ Section 14.2 On the Job
Answers will vary.

■ Section 14.3 Estimating on the Job
924 blocks are needed; 27.3 cubic feet of mortar; 51.3 hours

■ Section 14.3 Check Your Knowledge
1. (1) The walls do not require formwork.
(2) The blocks are relatively inexpensive.
(3) Work on a block foundation can be started and stopped as needed.
2. Type S mortar is particularly suitable for regions exposed to earthquakes.
3. A story pole is a board with markings 8" apart. It is used to gauge the top of the masonry for each course.
4. Parging is the process of spreading mortar or cement plaster over the block to form a cove. It is done to add water resistance to the wall.

■ Section 14.3 On the Job
Answers will vary.

■ Chapter 14 Review Questions
1. The minimum depth of a footing for an exterior wall is 12". However, in cold climates the footings should be far enough below grade to be protected from frost.
2. The reinforcement often consists of two lengths of ½" diameter steel rebar. The rebar must be positioned at least 3" above the bottom of the footing.
3. The keyway locks the foundation walls to the footing and makes it less likely that moisture will seep through. A 2×4 is often used to form it.
4. Medium-density overlay (MDO) has a smooth surface and can be reused many times. High-density overlay (HDO) offers the smoothest finish and the greatest number of reuses. Mill-oiled plywood has a sanded veneer surface that is coated with a release agent.
5. The two sides of each form are fastened together with clips or other ties. Thin metal rods called snap-ties are commonly used.
6. Damp proofing protects a wall against ordinary seepage, such as may occur after a rainstorm. Waterproofing is used where the soil drains poorly or where the water table is high. It is more effective than damp proofing.
7. A cold joint is created when fresh concrete is poured on top of or next to concrete that has already begun to cure.
8. Joints are tooled to make them neat and to seal them against water seepage.
9. Blocks are placed directly above one another, resulting in continuous vertical joints.
10. Mortar will stiffen on the mortar board either because of evaporation or hydration. When evaporation occurs, water can be added to the mortar to restore its workability. Mortar stiffened by hydration, however, should be discarded.

Instructional Plan

Concrete Flatwork

Focus & Planning

Objectives
- Identify the two types of foundation slabs.
- Understand foundation slab basics.
- Discuss various types of slab reinforcement.
- List the steps in finishing flatwork.

SAFETY FIRST
- ❏ Review the safety rules in the chapter.
- ❏ Make sure all necessary personal protective equipment is available and in good condition.
- ❏ Make sure all required tools and equipment are available and in good working order.

Instruction & Student Practice

Using the *Student Textbook*, have students:
- ❏ Read Section 15.1.
- ❏ Read Estimating Concrete for Flatwork and complete Section 15.1 Estimating on the Job.
- ❏ Complete Section 15.1 Check Your Knowledge and On the Job.
- ❏ Read Section 15.2.
- ❏ Complete Section 15.2 Check Your Knowledge and On the Job.

Using *Carpentry Applications*, assign:
- ❏ 15-1: Estimating Concrete
- ❏ 15-2: Estimating Concrete Using a Construction Calculator

Using *Carpentry Math*, assign:
- ❏ M15-1: Writing Ratios
- ❏ M15-2: Using Ratios to Mix Concrete
- ❏ M15-3: Estimating Concrete for Slabs

Review & Student Performance

- ❏ Assign Review Questions from the end of the chapter.
- ❏ Assign the Chapter 15 Test or create a test using the *ExamView® Pro Test Generator* on the *Instructor Productivity CD-ROM*.
- ❏ Create and administer a Unit 3 Posttest using the *ExamView® Pro Test Generator* on the *Instructor Productivity CD-ROM*.
- ❏ Use *Carpentry Applications* job sheets as desired for final performance assessment.

(Continued on next page)

Answers to Textbook Questions

■ Section 15.1 Estimating on the Job
7.4 cubic yards, 3.2 hours

■ Section 15.1 Check Your Knowledge
1. Concrete flatwork consists of flat, horizontal areas of concrete, usually 5" or less in thickness.
2. If fill is compacted in thicker layers, it may appear to be firm on the surface but it will not be uniformly firm.
3. The subbase helps to drain water that might accumulate under the slab.
4. Length (in feet) × width (in feet) × thickness (in feet) ÷ 27 = cubic yards.

■ Section 15.1 On the Job
Answers will vary

■ Section 15.2 Check Your Knowledge
1. It makes the concrete surface more uniform and eliminates high and low spots. A bull float also brings enough mortar to the surface of the slab to produce the desired finish.
2. Premature bullfloating brings an excess amount of fines and moisture to the surface. Among other problems, this causes fine hairline cracks (crazing) or a powdery material (dusting) at the surface.
3. A concrete slab is ready for jointing and edging when the sheen has left the surface and the concrete has started to stiffen.
4. They give the finisher something to kneel on as he/she works and helps avoid stepping on the fresh concrete.

■ Section 15.2 On the Job
Answers will vary.

■ Chapter 15 Review Questions
1. A monolithic slab consists of a footing and floor slab that are formed in one continuous pour. The perimeter of the slab is thicker than the main area and is strengthened with rebar at the edges. An independent slab on grade is poured between the foundation walls.
2. The reinforcement should be supported on special chairs made for this purpose.
3. This forces all moisture in the fresh concrete to evaporate through the exposed top surface of the material. This can cause shrinkage cracks to develop in the slab surface, as well as other problems.
4. Radon is a colorless, odorless radioactive gas given off by some soils and rocks.
5. A screed is a long, straight length of metal or wood that is used to strike off the concrete to a specified level.
6. Steps include screeding, bullfloating, edging, jointing, floating, and troweling.
7. The second troweling should be delayed until the concrete has become hard enough to make a ringing sound under the trowel.
8. Moist curing the slab for at least two days after it has been placed.
9. *Any three of the following*:
   - The water and aggregates should be kept as cool as possible prior to being mixed with the cement.
   - Forms, rebar, and the subgrade should be cooled by sprinkling them with water just before the concrete is placed.
   - In some cases, it may be prudent to place the concrete early in the morning, or even at night, to avoid hot temperatures.
   - Moist curing is particularly important under these conditions, and should be started as soon as possible.
10. *Concrete can be protected by any two of the following:*
    - Placing it in insulated forms
    - Covering it temporarily with insulation
    - Using either high-early strength, air-entrained, or low-slump concrete

Instructional Plan

CHAPTER 16: Wood As a Building Material

Focus & Planning

Objectives
- Identify flat-sawn and quarter-sawn boards.
- Explain how the moisture content of wood is controlled.
- Identify common defects in lumber.
- Recognize conditions that lead to the decay of lumber.
- Identify insects that can infest lumber.

SAFETY FIRST
- ❏ Review the safety rules in the chapter.
- ❏ Make sure all necessary personal protective equipment is available and in good condition.
- ❏ Make sure all required tools and equipment are available and in good working order.

Instruction & Student Practice

- ❏ Read Section 16.1.
- ❏ Complete Section 16.1 Check Your Knowledge and On the Job.
- ❏ Read Section 16.2.
- ❏ Complete Section 16.2 Check Your Knowledge and On the Job.

Using *Carpentry Applications*, assign:
- ❏ 16-1: Preventing Split Ends in Lumber
- ❏ 16-2: Learning Grades and Sizes

Using *Carpentry Math*, assign:
- ❏ M16-1: Reading Lumber Sizes
- ❏ M16-2: Calculating Board Feet and Lineal Feet

Carpenter's Tip

Hardwood sizes are often expressed in quarters of an inch. For example, 1 ¼" would be referred to as ⁵⁄₄, and 2 ½" would be ¹⁰⁄₄.

Using the *Safety Guidebook*, assign:
- ❏ 2-2: Personal Protective Equipment
- ❏ 3-8: Excavations
- ❏ 6-3: Hazardous Materials

Review & Student Performance

- ❏ Assign Review Questions from the end of the chapter.
- ❏ Assign the Chapter 16 Test or create a test using the *ExamView® Pro Test Generator* on the *Instructor Productivity CD-ROM*.
- ❏ Create and administer a Unit 4 Pretest using the *ExamView® Pro Test Generator* on the *Instructor Productivity CD-ROM*.
- ❏ Use *Carpentry Applications* job sheets as desired for final performance assessment.

(Continued on next page)

Answers to Textbook Questions

■ Section 16.1 Check Your Knowledge

1. Softwoods are those that come from coniferous trees, which produce their seeds in cones and have needlelike or scalelike leaves. Hardwoods are cut from broad-leaved, deciduous trees. A deciduous tree is one that sheds its leaves annually, during cold or very dry seasons.
2. It has a low tendency to warp, shrink, or swell; provides a more durable surface than flat-sawn lumber; and holds paints and finishes better.
3. KD lumber is kiln-dried lumber.
4. Lumber will pick up or lose moisture until it reaches a balance with the surrounding air.

■ Section 16.1 On the Job

Answers will vary.

■ Section 16.2 Check Your Knowledge

1. Fungi grow most rapidly at temperatures between 70°F and 85°F [21°C to 29°C].
2. The presence of moisture creates the best environment for decay.
3. The heartwood is the most decay resistant.
4. Subterranean termites thrive in moist, warm soil containing an abundant supply of food. In their search for additional food, they build tubes over foundation walls, in cracks, or on pipes or supports leading from the soil into the house. Dry-wood termites fly directly to the wood instead of building tunnels from the ground. They chew across the grain of the wood and excavate broad pockets, or chambers. These chambers are connected by narrow tunnels. The termites remain hidden in the wood and are seldom seen. Subterranean termites do the most damage.

■ Section 16.2 On the Job

The state is Florida. For the special building materials that can help prevent decay, answers will vary.

■ Chapter 16 Review Questions

1. *Any five of the following*:
   - Wood is a renewable resource.
   - Wood is light in weight and easily shipped, handled, and shaped.
   - Wood is easily fastened with nails, staples, bolts, connectors, screws, or glue.
   - Wooden buildings are easily altered or repaired.
   - Openings can be cut and additions made without difficulty.
   - Wood accepts decorative coatings such as paint and stains.
   - Wood is strong. Pound for pound, certain common framing woods are actually as strong and stiff as some types of steel.
   - Wood has low heat conductivity, which helps keep wooden buildings warm.
   - Wood resists rust, acids, saltwater, and other corrosive agents better than many other structural materials.
2. Most annual rings consist of a light band formed in the spring (early wood), and a dark band formed in the summer (late wood).
3. For flat-sawn lumber, a log is squared up lengthwise, and then sawed into boards. Growth rings run across a board's width. For quarter-sawn lumber, a log is first sawn lengthwise into quarters, and then boards are cut from the faces of each quarter.
4. When the cell walls of green wood have absorbed all the water they can hold, but there is no water in the cavities, the wood is at the fiber-saturation point.
5. Moisture content (MC) is expressed as a percentage of the weight of the wood.
6. Air drying is done outdoors. The rough lumber is stacked in layers separated by crosspieces called stickers. The wood remains stacked from one to three months and sometimes longer. In kiln drying, the lumber is also stacked in piles with stickers between layers. It is then placed in a kiln, an oven in which moisture, air, and temperature are carefully controlled.
7. Lumber should be supported off the ground and piled carefully with stickers between layers. It should be covered with waterproof material until used.
8. A bow is a flatwise deviation along the grain from a straight, true surface. A crook is an edgewise deviation from a straight, true surface. A cup is a flatwise deviation across the grain from a straight, true surface.
9. The two beetles are powderpost beetles and deathwatch beetles.
10. Preservative-treated wood is used outdoors or where it is in direct contact with concrete, masonry, or the ground.

Instructional Plan

Focus & Planning

Objectives
- Explain how the use of engineered lumber helps conserve wood resources.
- Tell how to store, handle, and install LVL-I-joists.
- Tell the differences among laminated-veneer lumber, glue-laminated lumber, laminated-strand lumber, and parallel-strand lumber.
- Recognize the connectors most often used with engineered lumber.

SAFETY FIRST
- ❑ Review the safety rules in the chapter.
- ❑ Make sure all necessary personal protective equipment is available and in good condition.
- ❑ Make sure all required tools and equipment are available and in good working order.

Instruction & Student Practice

Using the *Student Textbook*, have students:
- ❑ Read Section 17.1.
- ❑ Complete Section 17.1 Check Your Knowledge and On the Job.
- ❑ Read Section 17.2.
- ❑ Complete Section 17.2 Check Your Knowledge and On the Job.
- ❑ Read Section 17.3.
- ❑ Complete Section 17.3 Check Your Knowledge and On the Job.
- ❑ Read Section 17.4.
- ❑ Complete Section 17.4 Check Your Knowledge and On the Job.

Using *Carpentry Applications*, assign:
- ❑ 17-1: Installing LVL I-Joists
- ❑ 17-2: Installing Joist Hangers

Using *Carpentry Math*, assign:
- ❑ M17-1: Pricing Lumber

Using the *Safety Guidebook*, assign:
- ❑ 6-4: Bulk Lifting/Moving Operations

Review & Student Performance

- ❑ Assign Review Questions from the end of the chapter.
- ❑ Assign the Chapter 17 Test or create a test using the *ExamView® Pro Test Generator* on the *Instructor Productivity CD-ROM*.
- ❑ Use *Carpentry Applications* job sheets as desired for final performance assessment.

(Continued on next page)

Answers to Textbook Questions

■ Section 17.1 Check Your Knowledge

1. Plywood is cross-laminated. This means that the grain of each layer runs perpendicular to the grain of adjoining layers. The grain of every layer in LVL runs in the same direction. This is called parallel-lamination. This process produces a material that is even more uniform than would be found in a like thickness of material produced by cross-lamination. It also means that the end grain of each veneer layer is exposed only at the very ends of LVL products.
2. The easiest method is to use a radial-arm saw. To cut with a circular saw, place a wood block against the web and between the flanges to prevent the shoe of the saw from lodging against a flange during the cut.
3. The only time cutting a flange is permissible is when cutting the I-joist to length.
4. It is called a rim board.

■ Section 17.1 On the Job
Answers will vary.

■ Section 17.2 Check Your Knowledge

1. Glulam beams have significant resistance to fire. A glulam does not ignite easily, and it burns slowly if it does catch fire.
2. The best-quality material is used in the top and bottom layers of a glulam.
3. The glulam must not be notched or drilled unless these changes were allowed in the design.
4. Heavy-gauge metal framing connectors are often used to secure glulams.

■ Section 17.2 On the Job
Answers will vary.

■ Section 17.3 Check Your Knowledge

1. A finger joint is a closely spaced series of wedge-shaped cuts made in mating surfaces and glued.
2. It is consistently straight, can be sawn and nailed like solid lumber, makes use of short pieces of lumber that would be wasted otherwise and is available in longer lengths than standard lumber.
3. Finger-jointed lumber graded "certified exterior joints" can be used interchangeably with standard lumber of the same size, species, and grade. Lumber graded "certified glued joints" is suited to vertical use only.
4. Laminated-strand lumber is made from wood strands ranging from 0.03" to 0.05" thick, 1" wide, and approximately 12" long. Parallel-strand lumber is made from wood veneer ribbons (called strands) approximately 1" wide and up to 8' long.

■ Section 17.3 On the Job
Answers will vary.

■ Section 17.4 Check Your Knowledge

1. A joist hanger is a metal bracket used where floor or ceiling joists intersect another framing member, such as a beam.
2. They are typically installed with 10d common nails. However, using 16d common nails to fasten the connector to the intersecting support will improve the strength of the connection. Special joist-hanger nails may also be supplied by the manufacturer.
3. When I-joists will be attached to an intersecting I-joist, the joist hangers should be "backed." This involves placing backer blocks against the web, between the flanges of the supporting I-joist. This provides suitable nailing penetration.
4. The thickness of the zinc coating on a metal connector is indicated by a code. The standard zinc coating is G 60. This means that the zinc is 0.005" thick on each side of the steel.

■ Section 17.4 On the Job
Answers will vary.

■ Chapter 17 Review Questions

1. Engineered lumber includes manufactured products made of solid wood, wood veneer, wood pieces, or wood fibers bonded together with adhesives. It conserves wood resources by using wood that might otherwise be wasted.
2. It uses wood efficiently, has highly predictable performance, can be manufactured in a wide variety of sizes, and is free of the defects commonly found in solid lumber.
3. LVL lumber is made of layers of wood veneer bonded together with a waterproof adhesive (phenol resorcinol formaldehyde).
4. Laminated-veneer lumber leaves the mill at 8 percent moisture content.
5. Never drive nails sideways (parallel to the layers) into an I-joist flange. This tends to split the layers, reducing the strength of the product. Instead, nails should be driven into the flanges at a 45° angle.
6. Storing or carrying an I-joist on its side or allowing it to flex back and forth could damage the web. This would severely weaken the product.
7. Glulam beams can be used for garage door headers, patio door headers, carrying beams, window headers, and even exposed structural stair stringers. Glulam posts are also available.
8. Camber is a slight upward curve in a glulam.
9. Some manufacturers provide special nails, called joist-hanger nails, for use with their connectors. These nails have the same diameter as a 10d common nail, but they are shorter.
10. The most common mistake is to use too few nails. The connection depends on nails for shear strength. Thus, undernailing could cause it to fail when loads are placed on it.

Instructional Plan

Chapter 18: Engineered Panel Products

Focus & Planning

Objectives
- Identify the various types of engineered panels and their uses.
- Follow safety rules when handling or machining engineered panels.
- Explain the grading system for plywood.

SAFETY FIRST
- ☐ Review the safety rules in the chapter.
- ☐ Make sure all necessary personal protective equipment is available and in good condition.
- ☐ Make sure all required tools and equipment are available and in good working order.

Instruction & Student Practice

Using the *Student Textbook*, have students:
- ☐ Read Section 18.1.
- ☐ Complete Section 18.1 Check Your Knowledge and On the Job.
- ☐ Read Section 18.2.
- ☐ Complete Section 18.2 Check Your Knowledge and On the Job.

Using *Carpentry Applications*, assign:
- ☐ 18-1: Selecting Engineered Panel Products
- ☐ 18-2: Making a Plywood Ripping Guide

Using *Carpentry Math*, assign:
- ☐ M18-1: Estimating Sheathing
- ☐ M18-2: Estimating Sheathing Applications

Carpenter's Tip

Instead of laying plywood directly on sawhorses, try using a 2" sheet of Styrofoam plastic foam to protect the sawhorses.

Using the *Safety Guidebook*, assign:
- ☐ 2-1, 2-2. These relate to personal safety.
- ☐ 3-4, 6-2. These relate to hazardous materials.
- ☐ 5-13, 5-14. These relate to power saw safety.

Review & Student Performance

- ☐ Assign Review Questions from the end of the chapter.
- ☐ Assign the Chapter 18 Test or create a test using the *ExamView® Pro Test Generator* on the *Instructor Productivity CD-ROM*.
- ☐ Use *Carpentry Applications* job sheets as desired for final performance assessment.

(Continued on next page)

Answers to Textbook Questions

■ Section 18.1 Check Your Knowledge

1. *Any three*: Foundation forms, sheathing, soffits, underlayment, cabinets, built-ins.
2. All plywood is manufactured with an odd number of layers–usually 3, 5, or 7. The grain of each layer is placed at right angles to that of adjacent layers.
3. In veneer matching, veneer strips are aligned on a panel to achieve different visual effects.
4. A scrap piece of plywood is clamped beneath the area to be drilled. Drilling is done through the top piece and into the scrap. The scrap supports the edges of the hole.

■ Section 18.1 On the Job

1. The solid scrap will typically swell more than the plywood.
2. The solid scrap will usually show signs of warping as it dries; the plywood will show few signs, if any.

■ Section 18.2 Check Your Knowledge

1. Plywood is made from layers of wood veneer glued together. Composite panel products are made from pieces of wood mixed with adhesive and formed under pressure into uniform sheets.
2. Small amounts of wax (0.5 percent to 1 percent) reduce a panel's tendency to absorb moisture and make it more suitable for such uses as sheathing.
3. Fiberboard includes hardboard and medium-density fiberboard.
4. Fiber-cement board is used as siding.

■ Section 18.2 On the Job

Answers will vary.

■ Chapter 18 Review Questions

1. (1) They are engineered for the efficient use of wood resources. (2) They are manufactured using various types of natural or synthetic adhesives. (3) Their performance is highly predictable.
2. These are the outermost plies of a plywood panel.
3. Veneer is a very thin, pliable sheet of wood that is sawed, peeled, or sliced from a log.
4. *Any four of these identify its grade:* Its adhesive, veneer quality, wood species, construction, size, special characteristics, structural performance.
5. Predrilling a hole is generally necessary when starting screws into plywood by hand. Also, a hole should be pre-drilled when driving screws close to the edge of a plywood sheet. This prevents the edge from splitting.
6. When nails or screws alone will not hold plywood in place sufficiently, adhesives should also be used. The combination produces a particularly strong joint. This technique is called glue-nailing.
7. Many composite products incorporate chemicals and other additives. The MSDS will identify potential health and safety hazards associated with handling or machining the material. It will also suggest suitable precautions.
8. Oriented-strand board panels are made from reconstituted wood strands bonded with adhesive under heat and pressure. The strands are directionally oriented in layers that are perpendicular to each other. Panels will usually have three or five layers.
9. Engineered panels include plywood, oriented-strand board (OSB), fiberboard, particleboard, and fiber-cement board. Uses include foundation forms, underlayment, countertop substrates, sheathing, cabinetry, paneling, and siding.
10. Cutting fiber-cement board generates a great deal of very fine dust that cannot be contained by standard dust bags. Fiber cement contains silica, and breathing excessive amounts of silica dust can lead to an illness called silicosis. Always wear a dust mask when cutting fiber-cement products. If large quantities of dust are generated, a respirator may be necessary.

Instructional Plan

CHAPTER 19
Framing Methods

Focus & Planning

Objectives
- Name the stresses that structural wood must resist.
- Explain the difference between a live load and a dead load.
- Read a span table.
- Tell the differences between platform-frame construction and balloon-frame construction.
- List the advantages of structural insulated panels.

SAFETY FIRST
- ❏ Review the safety rules in the chapter.
- ❏ Make sure all necessary personal protective equipment is available and in good condition.
- ❏ Make sure all required tools and equipment are available and in good working order.

Instruction & Student Practice

Using the *Student Textbook*, have students:
- ❏ Read Section 19.1.
- ❏ Complete Section 19.1 Check Your Knowledge and On the Job.
- ❏ Read Section 19.2.
- ❏ Complete Section 19.2 Check Your Knowledge and On the Job.
- ❏ Read Section 19.3.
- ❏ Complete Section 19.3 Check Your Knowledge and On the Job.

Using *Carpentry Applications*, assign:
- ❏ 19-1: Building Outside Corners
- ❏ 19-2: Building Partition Corner Posts
- ❏ 19-3: Building a Trimmer and King Stud

Using *Carpentry Math*, assign:
- ❏ M19-1: Reading a Brace Table

Using the *Safety Guidebook*, assign:
- ❏ 6-4: Bulk Lifting/Moving Operations

Using the *Instructor Productivity CD-ROM*, show PowerPoint presentation:
- ❏ 19-1: Building Outside Corners
- ❏ 19-2: Building Partition Corner Posts

Review & Student Performance

- ❏ Assign Review Questions from the end of the chapter.
- ❏ Assign the Chapter 19 Test or create a test using the *ExamView® Pro Test Generator* on the *Instructor Productivity CD-ROM*.
- ❏ Use *Carpentry Applications* job sheets as desired for final performance assessment.

(Continued on next page)

Answers to Textbook Questions

■ Section 19.1 Check Your Knowledge

1. Design value tables show exactly how well each wood resists stresses caused by bending, tension, shear, and compression. Design values are used by architects and engineers when they wish to determine the exact performance of any part of a wood frame.
2. The modulus of elasticity is the ratio of the amount that wood will bend as compared to its load.
3. Span tables are available in building code books and in literature supplied by the major lumber trade associations.
4. Live load is anything not permanently attached to the building.

■ Section 19.2 Check Your Knowledge

1. Individual pieces include joists, studs, beams, and rafters. Wood panels (called sheathing) are fastened to the wood framing.
2. Sheathing consists of wood panels fastened to the framing to give it more strength and stiffness.
3. The basic types of conventional framing are platform-frame construction and balloon-frame construction.
4. In standard platform-frame construction, studs are commonly spaced 16" OC, but floor joists might be spaced at intervals of 12", 16", 19.2", or 24". With in-line framing, all joists, studs, and rafters are arranged on the same spacing. The spacing is usually 16" or 24" OC.

■ Section 19.2 On the Job

Answers will vary.

■ Section 19.3 Check Your Knowledge

1. A timber frame is a freestanding framing system that rests on a foundation. A timber frame is a type of post-and-beam frame.
2. Gable roofs are more susceptible to damage by high winds than hip roofs or flat roofs.
3. It is particularly important to provide this strength at the corners of a house.
4. Hold-down anchors can be installed at each corner of the house, or as required by local codes.

■ Section 19.3 On the Job

Answers will vary.

■ Chapter 19 Review Questions

1. Wood-frame houses often cost less than houses built using other structural systems. They are easily insulated, which reduces heating and air-conditioning costs. Wood framing is suitable for use with a wide variety of exteriors and architectural styles. A well-built and properly maintained wood-frame home is very durable.
2. Horizontal shear occurs.
3. This problem can be reduced simply by increasing the bearing area.
4. Deflection is the actual amount wood will bend when loaded. For example, deflection makes a floor feel "springy" when walked on.
5. A span table shows the maximum spacing allowed for joists or rafters based on the particular species, grade, and dimension of wood.
6. A dead load is the total weight of the building, including the structural frame itself and anything permanently attached to the building, such as wall coverings, cabinets, and roof shingles. Live loads are not permanently attached to the building. Examples include furniture, books, and people. Live loads are related to how the building is used.
7. In platform-frame construction each level of the house is constructed separately. In balloon-frame construction, the studs are continuous from the sill to the top plate of the second floor.
8. Post-and-beam framing calls for fewer but larger pieces of wood spaced farther apart than those used in conventional framing.
9. A structural insulated panel (SIP) consists of expanded polystyrene (EPS) foam insulation sandwiched between sheets of exterior plywood or oriented-strand board (OSB). There are several advantages to SIPs. The shell of the house can be erected very quickly. The house is very strong because there is wood sheathing on the inside as well as on the outside. The panels are energy efficient because they allow very little cold air to leak into the house.
10. A shear wall is designed to resist lateral forces such as those created by earthquakes or high winds.

Instructional Plan

CHAPTER 20: Floor Framing

Focus & Planning

Objectives
- Identify the basic floor-framing components and explain the purpose of each.
- Install posts and girders.
- Install sill plates and lay out basic joist spacing.
- Recognize cases where special framing details may be required, such as beneath a bearing wall.
- Lay a panel subfloor.
- Recognize other types of framing systems and products, such as those using trusses and girders.

SAFETY FIRST
- ❏ Review the safety rules in the chapter.
- ❏ Make sure all necessary personal protective equipment is available and in good condition.
- ❏ Make sure all required tools and equipment are available and in good working order.

Instruction & Student Practice

Using the *Student Textbook*, have students:
- ❏ Read Section 20.1.
- ❏ Complete Section 20.1 Check Your Knowledge and On the Job.
- ❏ Read Section 20.2.
- ❏ Read Estimating Floor Framing and complete Section 20.2 Estimating on the Job.
- ❏ Complete Section 20.2 Check Your Knowledge and On the Job.
- ❏ Read Section 20.3.
- ❏ Read Estimating Subflooring and complete Section 20.3 Estimating on the Job.
- ❏ Complete Section 20.3 Check Your Knowledge and On the Job.
- ❏ Read Section 20.4.
- ❏ Complete Section 20.4 Check Your Knowledge and On the Job.

Using *Carpentry Applications*, assign:
- ❏ 20-1: Installing Sill Plates
- ❏ 20-2: Laying Out Floor Joists
- ❏ 20-3: Installing Single-Layer Floor Panels

Using *Carpentry Math*, assign:
- ❏ M20-1: Installing Floor Girders
- ❏ M20-2: Estimating Floor Joists
- ❏ M20-3: Estimating Cross Bridging

Using the *Safety Guidebook*, assign:
- ❏ 5-12: Powder-Actuated Tools
- ❏ 5-25: Power Nails and Staplers

Using the *Instructor Productivity CD-ROM*, show PowerPoint presentation:
- ❏ 20-1: Installing Sill Plates
- ❏ 20-2: Laying Out Floor Joists

Review & Student Performance

- ❏ Assign Review Questions from the end of the chapter.
- ❏ Assign the Chapter 20 Test or create a test using the *ExamView® Pro Test Generator* on the *Instructor Productivity CD-ROM*.
- ❏ Use *Carpentry Applications* job sheets as desired for final performance assessment.

(Continued on next page)

Answers to Textbook Questions

■ Section 20.1 Check Your Knowledge

1. A girder is a large principal horizontal member used to support floor joists. One or more posts provide intermediate support for a girder. The girder ends are supported by the foundation walls.
2. A Lally column is a steel post.
3. Face-nail each layer with 10d nails as follows: Stagger nails 32" OC at top and bottom. Nail two or three times at the end of every board, including splices. Arrange joints so that they are staggered.
4. Wood plates must be fastened to a steel girder so that floor joists can be toenailed to them later on.

■ Section 20.1 On the Job

Answers will vary, but should include details about the type, spacing, and dimension of fasteners. It should also identify any metal connecting brackets, bolts, or screws by the product number or type and size.

■ Section 20.2 Estimating on the Job

$48 \times .75 = 36$
$36 + 1 = 37$
$37 \times 4 = 148$
$148 \times 12 = 1776$
1776 lineal feet will be required (148 joists).

■ Section 20.2 Check Your Knowledge

1. The sill plate provides a smooth bearing surface for the floor joists and serves as a connection between the foundation wall and the floor system.
2. Use a ⅝" spade bit.
3. The crown is the outermost curve of a bowed joist. If the crown is placed on top, a bowed joist will tend to straighten out when subfloor and normal floor loads are applied to it.
4. A web stiffener is a small block of plywood or OSB sheathing that is nailed to the web of an I-joist. It can be added to both sides of the I-joist to improve the fit of metal framing anchors. Where a wood I-joist runs continuously over a support, web stiffeners should be added to both sides to improve the bearing characteristics.

■ Section 20.2 On the Job

Answers will vary.

■ Section 20.3 Estimating on the Job

$48 \times 22 = 1,056$
$1,056 \times 2 = 2,112$
$2,112 \div 32 = 66$
66 4×8 sheets will be required.

■ Section 20.3 Check Your Knowledge

1. Apply adhesive just before a panel is placed.
2. Common nails, ring-shank nails, screw-shank nails, and screws are used to attach sheathing.
3. Underlayment is a panel product that is laid over the subfloor. It covers any minor construction damage to the subfloor and provides a smooth substrate for the finish flooring. It is used beneath finish floors such as sheet vinyl.
4. Use a 10d box nail, approximately ⅛" in diameter.

■ Section 20.3 On the Job

Answers will vary.

■ Section 20.4 Check Your Knowledge

1. The three parts are the chords, the webs, and the connector plates.
2. Posts are supported by a row of concrete piers resting atop concrete footings.
3. The bearing posts must be cut to length accurately in order to provide a level floor.
4. They are roughed in before the subfloor is laid.

■ Section 20.4 On the Job

Answers will vary.

■ Chapter 20 Review Questions

1. The basic components are posts, girders, sill plates, joists and trusses, and subflooring.
2. A girder is placed halfway between the longest foundation walls and parallel to them.
3. Check that the foundation is level and square.
4. The edges must fall along the centerline of a joist.
5. Floor frames are bridged in order to stiffen the floor system, prevent unequal deflection (bending) of the joists, prevent twisting, and to enable an overloaded joist to receive some assistance from the joists on either side of it.
6. Joists should be doubled under each bearing wall that is parallel to the joists. If the wall will contain plumbing pipes or heating ducts, the joists can be separated by blocking.
7. Tail joists are floor joists that have been interrupted by a header. They are often supported by metal framing connectors nailed to the header.
8. Doubled joists provide a nailing surface for the interior wall finish.
9. Sheathing lends bracing strength to the building. It provides a solid base for the finish floor. By acting as a barrier to cold and dampness, sheathing helps keep the building warmer and drier in winter. In addition, it provides a safe working surface for building the house.
10. Concrete piers and footings support the posts.

Instructional Plan

Focus & Planning

Objectives
- Identify wall-framing members.
- Lay out a wall.
- Assemble and erect a wall.
- Identify situations that require special framing.
- Apply sheathing.
- Estimate materials for wall framing and sheathing.

SAFETY FIRST
- ❏ Review the safety rules in the chapter.
- ❏ Make sure all necessary personal protective equipment is available and in good condition.
- ❏ Make sure all required tools and equipment are available and in good working order.

Instruction & Student Practice

Using the *Student Textbook*, have students:
- ❏ Read Section 21.1.
- ❏ Read Estimating Studs, Plates, and Headers.
- ❏ Complete Section 21.1 Estimating on the Job.
- ❏ Complete Section 21.1 Check Your Knowledge and On the Job.
- ❏ Read Section 21.2.
- ❏ Complete Section 21.2 Check Your Knowledge and On the Job.
- ❏ Read Section 21.3.
- ❏ Complete Section 21.3 Check Your Knowledge and On the Job.
- ❏ Read Section 21.4.
- ❏ Complete Section 21.4 Check Your Knowledge and On the Job.
- ❏ Read Section 21.5.
- ❏ Read Estimating Number of Sheets Needed and complete Section 21.5 Estimating on the Job.
- ❏ Complete Section 21.5 Check Your Knowledge and On the Job.

Using *Carpentry Applications*, assign:
- ❏ 21-1: Laying Out Studs for a Wall
- ❏ 21-2: Installing a Let-in Diagonal Brace

Using *Carpentry Math*, assign:
- ❏ M21-1: Estimating Wall Framing Materials
- ❏ M21-2: Estimating the Length of Diagonals

Using the *Safety Guidebook*, assign:
- ❏ 5-13 through 5-17. These relate to safe use of various types of power saws.
- ❏ 5-25: Power Nailers and Staplers

Using the *Instructor Productivity CD-ROM*, show PowerPoint presentation:
- ❏ 21-1: Laying Out Studs for a Wall

Review & Student Performance

- ❏ Assign Review Questions from the end of the chapter.
- ❏ Assign the Chapter 21 Test or create a test using the *ExamView® Pro Test Generator* on the *Instructor Productivity CD-ROM*.
- ❏ Use *Carpentry Applications* job sheets as desired for final performance assessment.

(Continued on next page)

Answers to Textbook Questions

■ Section 21.1 Check Your Knowledge
1. It is 16" OC.
2. The sole plate is another term.
3. It supports structural loads and transfers them to the trimmer studs.
4. Without cripple studs: (1) The window would not be supported properly. (2) The wall would be weakened in that area because the sheathing would not be attached to the framing. (3) There would be no way to support interior wall surfaces such as drywall.

■ Section 21.1 On the Job
Sixty-two studs are needed.

■ Section 21.2 Check Your Knowledge
1. It is used to check whether intersecting walls are exactly 90° to each other.
2. Plate layout is the process of marking the location of studs, doors, and windows on the top and bottom plates.
3. A by-wall runs from the outside edge of the subfloor to the outside edge of the opposite end of the subfloor. Butt-walls fit between the by-walls.
4. A rough opening is the space into which a door or window will fit. It allows room for the door or window and its frame and allows space for leveling and plumbing the frame.

■ Section 21.2 On the Job
Answers will vary, but should reflect students' understanding of the procedure for determining whether walls are square.

■ Section 21.3 Check Your Knowledge
1. Story pole serves as standard reference for location and size of window headers, sills, door headers, and heights of various openings above subfloor.
2. The header may settle, causing cracks in the wall finish and making the door or window fit improperly.
3. The length of the opening it must span determines the depth of a header.
4. Two 16d nails are used.

■ Section 21.3 On the Job
The top of the wall might be out of plumb by as much as ¼".

■ Section 21.4 Check Your Knowledge
1. It is a wall that has been engineered to withstand unusual stresses. Shear walls are often encountered in areas where earthquakes and severe storms are common.
2. *Any three*: The types of special framing in a bathroom are framing to support the edge of a bathtub, for wall-hung sinks and toilets, for grab rails, for the shower-arm fitting, and for plumbing access doors. Special framing might also be needed to accommodate plumbing vents.
3. The cabinet soffit is commonly used where prefabricated upper cabinets do not extend all the way to the ceiling.
4. It slows the spread of fire through wall cavities.

■ Section 21.4 On the Job
Answers will vary.

■ Section 21.5 Estimating on the Job
Fifty-nine sheets will be needed.

■ Section 21.5 Check Your Knowledge
1. The most common sheathings are square-edged 4×8 panels made of plywood or OSB.
2. With ½" sheathing panels, 6d nails should be spaced 12" OC. The size of nail might change if the thickness of the sheathing changed.
3. Blocking should be installed.
4. This provides additional rigidity to the structure.

■ Section 21.5 On the Job
Thirty-four sheets will be required; *cost will vary*.

■ Chapter 21 Review Questions
1. If a wall supports weight from portions of the house above, such as the roof, it is a bearing wall. If a wall supports only its own weight and that of the wall covering, it is a nonbearing wall.
2. The requirements for wall-framing lumber are stiffness, good nail-holding capability, freedom from warpage and twist, and ease of workability.
3. The primary wall-framing members are studs and plates. Members that connect to them are headers, sills, cripple studs, and trimmer studs.
4. Three types of plates are bottom (sole) plate, top plate, and double (rafter) plate.
5. The two main steps are (1) mark the location of walls on the subfloor and (2) mark the location of the studs, windows, and doors on the wall plates.
6. If the edges are not flush, the sheathing won't fit correctly.
7. *Any three*: Special framing is needed for unusual architectural features, to provide openings for plumbing vents and fixtures, to provide openings for heating ducts, to add support for heavy items, to add blocking that supports the edges of interior wall coverings, and for fire safety.
8. It strengthens and braces the wall framing, forms a solid nailing base for the siding, helps to seal the house by reducing air infiltration, and ties wall framing to floor framing.
9. Estimate one stud per lineal foot of wall.
10. You should multiply the height by the width.

Instructional Plan

Chapter 22: Basic Roof Framing

Focus & Planning

Objectives
- Identify the basic roof styles.
- Understand the basic terms relating to roof-frame carpentry.
- Develop framing plans for a gable roof, hip roof, and variations that include valleys.
- Lay out a common rafter, using at least one of the four basic methods.
- Lay out ceiling joists.
- Recognize when special ceiling framing may be required.

SAFETY FIRST
- ❑ Review the safety rules in the chapter.
- ❑ Make sure all necessary personal protective equipment is available and in good condition.
- ❑ Make sure all required tools and equipment are available and in good working order.

Instruction & Student Practice

Using the *Student Textbook*, have students:
- ❑ Read Section 22.1.
- ❑ Complete Section 22.1 Check Your Knowledge and On the Job.
- ❑ Read Section 22.2.
- ❑ Complete Section 22.2 Check Your Knowledge and On the Job.
- ❑ Read Section 22.3.
- ❑ Complete Section 22.3 Check Your Knowledge and On the Job.

Using *Carpentry Applications*, assign:
- ❑ 22-1: Basic Roof Calculations
- ❑ 22-2: Laying Out a Common Rafter for a Gable Roof
- ❑ 22-3: Laying Out a Common Rafter Using a Construction Calculator

Using *Carpentry Math*, assign:
- ❑ M22-1: Calculating Pitch
- ❑ M22-2: Calculating Slope
- ❑ M22-3: Distinguishing Pitch from Slope

Using the *Safety Guidebook*, assign:
- ❑ 2-1: Musculoskeletal Disorders
- ❑ 4-3: Working on Roofs
- ❑ 5-4, 5-25. These relate to nailing safely.

Using the *Instructor Productivity CD-ROM*, show PowerPoint presentation:
- ❑ 22-1: Basic Roof Calculations
- ❑ 22-2: Laying Out a Common Rafter for a Gable Roof

Review & Student Performance

- ❑ Assign Review Questions from the end of the chapter.
- ❑ Assign the Chapter 22 Test or create a test using the *ExamView® Pro Test Generator* on the *Instructor Productivity CD-ROM*.
- ❑ Use *Carpentry Applications* job sheets as desired for final performance assessment.

(Continued on next page)

Answers to Textbook Questions

■ Section 22.1 Check Your Knowledge

1. Gable roof, hip roof, flat roof, and shed roof are the four basic roof styles.
2. The gambrel roof is a variation of the gable roof.
3. Any line that is vertical when a rafter is in its proper position is called a *plumb line*. Any line that is horizontal when a rafter is in its proper position is called a *level line*.
4. A roof framing plan is made to determine the kinds of rafters that will be needed for framing. If the plan is drawn to scale, the exact number of each kind of rafter can also be determined.

■ Section 22.1 On the Job

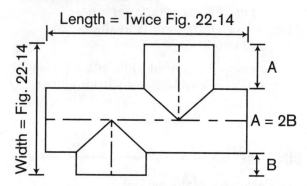

■ Section 22.2 Check Your Knowledge

1. The ceiling joists act as a tie and prevent the rafters from pushing out on the exterior walls.
2. The Pythagorean theorem states that the square of the hypotenuse of a right triangle is equal to the sum of the squares of the other two sides. Also, the length of the hypotenuse will be the square root of the sum of the squares of the other two sides.
3. The *heel cut* bears against the side of the plate. The *seat cut* bears on the top of the plate.
4. It is very important not to overcut when making the bird's mouth. This will weaken the rafter.

■ Section 22.2 On the Job

The rafter's length would be 13'-7 ¼".

■ Section 22.3 Check Your Knowledge

1. The size of the ceiling joists is determined by the distance they must span and the load they must carry. The species and grade of wood are also factors to be considered.
2. Ceiling joists are usually placed across the width of the building and parallel to the rafters.
3. Codes require that any framing, including ceiling framing, be kept at least 2" from the front and sides of masonry chimneys.
4. Ceiling joists are fastened to the side of the girder with joist hangers.

■ Section 22.3 On the Job

Answers will vary.

■ Chapter 22 Review Questions

1. The mansard is a variation of the hip roof.
2. A basic roof consists of rafters, collar ties, ceiling joists, and a ridge board.
3. The four basic types of rafters are common rafters, hip rafters, valley rafters, and jack rafters.
4. It is the distance between the outside edges of the double top plates, measured at right angles to the ridge board.
5. A plumb cut line is drawn using the framing square as a guide. The unit run (12" mark) on the blade of the square is aligned with the edge of the rafter. The unit rise on the tongue of the square (corresponding to the pitch of the roof) is aligned on the same edge of the rafter. The plumb cut line is then drawn along the edge of the tongue.
6. The unit length is found on the rafter table of the framing square.
7. The methods are the Pythagorean-theorem method, unit-length method, step-off method, and calculator method.
8. Carpenters using conventional roof framing methods can gang-cut rafters.
9. The ends of the joists that rest on the exterior wall plates next to the rafters will usually project above the top edge of the rafter. These ends must be cut off at an angle that is equal to the roof pitch.
10. Openings are required for chimneys and attic access.

Instructional Plan

ack Rafters

Focus & Planning

Objectives
- Lay out a hip rafter for a given roof.
- Lay out a valley rafter for a given roof.
- Lay out a jack rafter for a given roof.
- Explain why the intersection of two roofs makes framing more complex.

SAFETY FIRST
- ❏ Review the safety rules in the chapter.
- ❏ Make sure all necessary personal protective equipment is available and in good condition.
- ❏ Make sure all required tools and equipment are available and in good working order.

Instruction & Student Practice

Using the *Student Textbook*, have students:
- ❏ Read Section 23.1.
- ❏ Complete Section 23.1 Check Your Knowledge and On the Job.
- ❏ Read Section 23.2.
- ❏ Complete Section 23.2 Check Your Knowledge and On the Job.
- ❏ Read Section 23.3.
- ❏ Complete Section 23.3 Check Your Knowledge and On the Job.

Using *Carpentry Applications*, assign:
- ❏ 23-1: Laying Out a Hip Rafter
- ❏ 23-2: Laying Out a Hip Rafter Using a Construction Calculator
- ❏ 23-3: Laying Out a Hip Jack Rafter

Using *Carpentry Math*, assign:
- ❏ M23-1: Reading Rafter Tables

Using the *Safety Guidebook*, assign:
- ❏ 5-13: Circular Saws
- ❏ 5-25: Miter Saws

Using the *Instructor Productivity CD-ROM*, show PowerPoint presentation:
- ❏ 23-1: Laying Out a Hip Rafter
- ❏ 23-3: Laying Out Hip Jack Rafter

Review & Student Performance

- ❏ Assign Review Questions from the end of the chapter.
- ❏ Assign the Chapter 23 Test or create a test using the *ExamView® Pro Test Generator* on the *Instructor Productivity CD-ROM*.
- ❏ Use *Carpentry Applications* job sheets as desired for final performance assessment.

(Continued on next page)

Answers to Textbook Questions

■ Section 23.1 Check Your Knowledge

1. A hip rafter forms a raised area; a valley rafter forms a depression.
2. The unit run of a hip rafter is the hypotenuse of a right triangle with the shorter sides each equal to the unit run of a common rafter. Using the Pythagorean theorem ($12^2 + 12^2$), the unit run of the hip rafter is 16.97, which is rounded to 17.
3. The shortening allowance is one-half the 45° thickness of the ridge board.
4. Dropping the hip means to deepen the bird's mouth so as to bring the top edge of the hip rafter level with the upper ends of the jacks.

■ Section 23.1 On the Job

Answers will vary.

■ Section 23.2 Check Your Knowledge

1. An equal-span roof is a roof in which the span of the addition or dormer roof is the same as the span of the main roof.
2. They will intersect and be at the same height.
3. The dormer's ridge board is fastened to a header set between a pair of doubled common rafters in the main roof.
4. The total run of a valley rafter is the hypotenuse of a right triangle the shorter sides of which are each equal to the run of a common rafter in the dormer.

■ Section 23.2 On the Job

Answers will vary.

■ Section 23.3 Check Your Knowledge

1. A jack rafter is a shortened common rafter. It may be framed to a hip rafter, a valley rafter, or both.
2. A valley jack rafter extends from a valley rafter to a ridge board.
3. Lay out a framing plan. By studying the plan you can figure the total lengths of the valley jacks and cripple jacks.
4. Rather than lay out and mark each jack rafter individually, a pattern can be used to save time. When all the rafters have been cut, the pattern is used as a part of the roof frame.

■ Section 23.3 On the Job

The shortest one is 17.89" (approx. 17 ⅞").
The next longest one is 35.78" (approx. 35 ¾").
$6^2 + 12^2 =$
$36 + 144 = 180$
square root of 180 = 13.41641, rounded to 13.42

13.42 = unit length

$$\frac{12}{13.42} = \frac{16}{x}$$

$$\frac{214.72}{12} = 17.89 \text{ common difference}$$

Therefore, the shortest jack will be 17.89" long, or 17 ⅞". The next longest jack will be:
$2 \times 17.89 = 35.78$" long, or 35 ¾"

■ Chapter 23 Review Questions

1. A hip rafter is a roof member that forms a raised area, or "hip," in the roof, usually extending from the corner of the building diagonally to the ridge board.
2. A valley rafter forms a depression in the roof. Like the hip rafter, it extends diagonally from plate to ridge board.
3. The type called for is a hip roof with an intersecting addition.
4. A cut made where the end of a hip rafter meets the ridge board or the ends of common rafters at an angle.
5. The run of the hip rafter overhang is the hypotenuse of a right triangle the shorter sides of which are equal to the run of a common rafter overhang.
6. Backing the hip means to bevel the upper edge of the hip rafter.
7. The addition ridge board is at a lower level.
8. The shortening allowance is one-half the 45° thickness of the inside member in the doubled common rafter.
9. A hip jack rafter extends from a hip rafter to a rafter plate.
10. The exact method for framing the addition's ridge must be determined from among several choices. The length of the addition's ridge must be calculated. When two gable roofs meet, valley rafters are required, along with valley jack rafters.

Instructional Plan

CHAPTER 24: Roof Assembly and Sheathing

Focus & Planning

Objectives
- Identify the two basic types of ridges.
- Calculate ridge length.
- Create the ridge layout for gable roofs, hip roofs, addition roofs, and dormers.
- Lay out the locations of common rafters on a gable or hip roof.
- Identify where special framing details are required.
- Understand the basic requirements for the placement and nailing of panel roof sheathing.

SAFETY FIRST
- ❏ Review the safety rules in the chapter.
- ❏ Make sure all necessary personal protective equipment is available and in good condition.
- ❏ Make sure all required tools and equipment are available and in good working order.
- ❏ When carpenters are nailing rafters to a ridge board or ridge beam, they work high above the subfloor. Safety must always be a primary concern because falls are a constant risk.

Instruction & Student Practice

Using the *Student Textbook*, have students:
- ❏ Read Section 24.1.
- ❏ Complete Section 24.1 Check Your Knowledge and On the Job.
- ❏ Read Section 24.2.
- ❏ Read Estimating Roofing Materials and complete Section 24.2 Estimating on the Job.
- ❏ Complete Section 24.2 Check Your Knowledge and On the Job.
- ❏ Read Section 24.3.
- ❏ Complete Section 24.3 Check Your Knowledge and On the Job.
- ❏ Read Section 24.4.
- ❏ Read Estimating Sheathing and complete Section 24.4 Estimating on the Job.
- ❏ Complete Section 24.4 Check Your Knowledge and On the Job.

Using *Carpentry Applications*, assign:
- ❏ 24-1: Laying Out Studs on Gable Ends Using a Square
- ❏ 24-2: Laying Out Studs on a Gable End Using the Construction Calculator

Using *Carpentry Math*, assign:
- ❏ M24-1: Calculating Rafter Line Length

Using the *Safety Guidebook*, assign:
- ❏ 4-3: Working on Roofs

Review & Student Performance

- ❏ Assign Review Questions from the end of the chapter.
- ❏ Assign the Chapter 24 Test or create a test using the *ExamView® Pro Test Generator* on the *Instructor Productivity CD-ROM*.
- ❏ Use *Carpentry Applications* job sheets as desired for final performance assessment.

(Continued on next page)

Answers to Textbook Questions

■ Section 24.1 Check Your Knowledge

1. A ridge board is a nonstructural member that serves as a bearing surface for opposing pairs of rafters. The rafters hold the ridge board in place. A ridge beam supports the upper ends of the rafters and is itself supported by posts.
2. The extra width of the ridge board ensures that the ends of the rafters will bear fully on it.
3. It is shorter because it does not extend to the ends of the building.
4. For an equal-span addition, the length of the ridge board is equal to the distance that the addition projects beyond the building, plus one-half the span of the building, minus the shortening allowance at the main roof ridge.

■ Section 24.1 On the Job
Answers will vary.

■ Section 24.2 Estimating on the Job
1,219.4 board feet; 1 pound of nails (10.97)

■ Section 24.2 Check Your Knowledge

1. Lay out the rafter locations on the top plates first. Transfer the locations to the ridge board by laying the ridge board on edge against a top plate and matching marks.
2. Nailing at the plate first prevents the rafter from slipping out of position as the ridge is being installed.
3. Ceiling-joist ends are nailed to adjacent rafters with four 10d nails, two to each side. Metal brackets may also be used to attach the rafters to the plate.
4. If all the hip jacks are installed on one side of the hip, this pushes the hip out of alignment and causes it to bow.

■ Section 24.2 On the Job
Answers will vary.

■ Section 24.3 Check Your Knowledge

1. A collar tie is a horizontal framing member that prevents opposing rafter pairs from spreading apart. It also prevents the rafters from bowing inwards. In a finished attic, the collar ties may support the ceiling surfaces.
2. A system of purlins and braces can be used to support rafters that must span a longer distance.
3. It is the difference in length between one gable-end stud and the next, a measure based on the pitch of the roof.
4. Dormers are framed when all of the common rafters are in place.

■ Section 24.3 On the Job
Answers will vary.

■ Section 24.4 Estimating on the Job
46.68 panels

■ Section 24.4 Check Your Knowledge

1. The number in front of the slash indicates the maximum spacing of supports when the panel is used for roof sheathing. Thus this panel could be installed over rafters spaced 32" OC.
2. Nails should be spaced no more than 6" OC at supported panel ends and edges, and 12" apart at intermediate supports. Nails should be placed approximately ⅜" in from panel ends and edges.
3. It promotes ventilation around the shingles and allows them to dry out evenly.
4. Where chimney openings occur in the roof structure, the roof sheathing should have a clearance of ¾" from the finished masonry on all sides.

■ Section 24.4 On the Job
Answers will vary.

■ Chapter 24 Review Questions

1. The two types are nonstructural ridge boards and structural ridge beams.
2. The theoretical length of the ridge board (or ridge beam) is equal to the length of the building, measured to the outside edge of the wall framing. The actual length of the ridge board includes any overhang.
3. The shortening allowance accounts for the thickness of the main roof ridge board when determining the length of an intersecting ridge.
4. This will ensure that the rafters butt to the ridge board directly opposite each other.
5. (1) The number of rafters may be counted directly from the roof framing plan. (2) For rafters on 16" centers, take three-fourths of the building's length in feet, add one, then double this figure.
6. *Any five:* Collar ties, purlins and braces, gable ends, dormers, roof openings, chimney saddles.
7. The ends must be cut to prevent the tie from obstructing roof sheathing. To figure the angle, lay out the end cuts with a framing square set to the pitch of the roof.
8. The formula is
   16" ÷ 12" = 1 ⅓"
   1 ⅓" × 6" = 8"
9. Unsupported edge joints can be strengthened with metal panel clips that tie the joints together. These joints can also be supported by wood blocks.
10. The space between roof sheathing panels should be ⅛" or as recommended by the manufacturer. This is done because panels will shrink or swell slightly as their moisture content changes. If panels are butted tightly during installation, they may buckle as they expand.

Instructional Plan

Focus & Planning

Objectives
- Name the three basic parts of a roof truss.
- Handle roof trusses properly.
- Properly store roof trusses on the job site.
- Install roof trusses.

SAFETY FIRST
- ❑ Review the safety rules in the chapter.
- ❑ Make sure all necessary personal protective equipment is available and in good condition.
- ❑ Make sure all required tools and equipment are available and in good working order.

Instruction & Student Practice

Using the *Student Textbook*, have students:
- ❑ Read Section 25.1.
- ❑ Complete Section 25.1 Check Your Knowledge and On the Job.
- ❑ Read Section 25.2.
- ❑ Complete Section 25.2 Check Your Knowledge and On the Job.

Using *Carpentry Applications*, assign:
- ❑ 25-1: Ordering Trusses
- ❑ 25-2: Building a Truss

Using *Carpentry Math*, assign:
- ❑ M25-1: Estimating Rafters or Trusses

Using the *Safety Guidebook*, assign:
- ❑ 6-4: Bulk Lifting/Moving Operations

Using the *Instructor Productivity CD-ROM*, show PowerPoint presentation:
- ❑ 25-1: Ordering Trusses

Review & Student Performance

- ❑ Assign Review Questions from the end of the chapter.
- ❑ Assign the Chapter 25 Test or create a test using the *ExamView® Pro Test Generator* on the *Instructor Productivity CD-ROM*.
- ❑ Create and administer a Unit 4 Posttest using the *ExamView® Pro Test Generator* on the *Instructor Productivity CD-ROM*.
- ❑ Use *Carpentry Applications* job sheets as desired for final performance assessment.

(Continued on next page)

Answers to Textbook Questions

■ Section 25.1 Check Your Knowledge

1. A chord is the top or bottom outer member of a truss.
2. The most commonly used trusses are the king-post, Fink (also called W-truss), and scissors truss.
3. The design of a truss depends not only on the loads it must carry, but also on the weight and slope of the roof itself. Generally, the flatter the slope, the greater the stresses. Flatter slopes therefore require larger members and stronger connections in roof trusses.
4. Generally, the nominal span is the length of the bottom chord.

■ Section 25.1 On the Job

1. Answers will vary.
2. Answers will vary.
3. Answers will vary.
4. Answers will vary.

■ Section 25.2 Check Your Knowledge

1. Avoid placing unusual stresses on them. They should be lifted and stored upright. If they must be handled or stored in a flat position, enough support should be used along their length to minimize bending. Never support the trusses only at the center or only at each end when they are in a flat position.
2. Completed trusses can be raised into place by hand or by crane. Using a crane is the preferred method.
3. Another worker should be at the roof level to brace the trusses.
4. The gable-end truss should be braced with lumber standoffs anchored to stakes driven into the ground. As each additional truss is put into place, it should be braced temporarily to adjacent trusses with a length of nominal 2" lumber. The lumber should be secured diagonally to the top chord of each truss with two 16d nails. Lateral bracing may also be required. Temporary bracing can be removed as the roof sheathing is installed.

■ Section 25.2 On the Job

Answers will vary.

■ Chapter 25 Review Questions

1. Any three of the following:
   - The roof truss is capable of supporting loads over long spans.
   - Trusses save money on materials and on-site labor.
   - Roof trusses can eliminate the need for interior bearing partitions because trusses are self-supporting.
   - Trusses can be erected quickly.
   - Trusses allow increased flexibility for interior planning, and partitions can be placed without regard to structural requirements.
   - Roof trusses come in a wide variety of shapes to solve nearly any problem.
2. Chords are the outer elements of the truss. Webs create a rigid assembly within the chords. Chords and webs are typically connected at the joints by rectangular connector plates made of metal.
3. Trusses can also be built from metal and timber.
4. Trusses are commonly designed for 24" OC spacing.
5. They should be supported above the ground to protect them from dampness and water. A tarp should protect them from rain. Trusses must be properly supported if it is necessary to store them flat. If the trusses are stored vertically, it is important to prevent them from tipping and possibly injuring nearby workers.
6. Stand well to the side while cutting these bands. Make sure that others are standing well away.
7. Not only does this help to maintain precise spacing, but it also prevents them from tipping over like dominoes.
8. Continuous lateral bracing consists of 2×4 or wider stock that is nailed to the webs or lower chord of each truss. The exact location of the lateral bracing is usually specified in the truss design.
9. Temporary bracing can be removed once the permanent bracing is in place and after the sheathing is applied.
10. A better system of fastening trusses involves the use of a metal bracket. The brackets are nailed to the top and sides of the wall plate and to the lower chord of the truss.

Instructional Plan

CHAPTER 26: Steel Framing Basics

Focus & Planning

Objectives
- Describe the performance and prescriptive methods for steel framing design.
- Describe the three types of steel frame construction.
- Identify tools used in steel framing.
- Identify the head styles of steel-framing screws.
- Tell the difference between welding and clinching.

SAFETY FIRST
- ❏ Review the safety rules in the chapter.
- ❏ Make sure all necessary personal protective equipment is available and in good condition.
- ❏ Make sure all required tools and equipment are available and in good working order.

Instruction & Student Practice

Using the *Student Textbook*, have students:
- ❏ Read Section 26.1.
- ❏ Complete Section 26.1 Check Your Knowledge and On the Job.
- ❏ Read Section 26.2.
- ❏ Complete Section 26.2 Check Your Knowledge and On the Job.
- ❏ Read Section 26.3.
- ❏ Complete Section 26.3 Check Your Knowledge and On the Job.

Using *Carpentry Applications*, assign:
- ❏ 26-1: Getting Started in Steel Framing
- ❏ 26-2: Framing an Opening with Steel

Using *Carpentry Math*, assign:
- ❏ M26-1: Using a Construction Calculator to Convert Measurements
- ❏ M26-2: Using a Construction Calculator to Estimate Materials
- ❏ M26-3: Using a Construction Calculator to Find Circumference and Area
- ❏ M26:4: Using a Construction Calculator to Determine Layouts

Using the *Safety Guidebook*, assign:
- ❏ 1-5, 2-2, 2-3. These relate to personal safety.
- ❏ 5-1, 5-7, 5-8, 5-10, 5-11, 5-13, 5-25, 5-26. These relate to the use of various types of tools.

Using the *Instructor Productivity CD-ROM*, show PowerPoint presentation:
- ❏ 26-1: Getting Started in Steel Framing

Review & Student Performance

- ❏ Assign Review Questions from the end of the chapter.
- ❏ Assign the Chapter 26 Test or create a test using the *ExamView® Pro Test Generator* on the *Instructor Productivity CD-ROM*.
- ❏ Create and administer a Unit 26 Pretest using the *ExamView® Pro Test Generator* on the *Instructor Productivity CD-ROM*.
- ❏ Use *Carpentry Applications* job sheets as desired for final performance assessment.

(Continued on next page)

Answers to Textbook Questions

Section 26.1 Check Your Knowledge

1. Cold-formed steel is sheet steel that is bent and formed without using heat.
2. Architects and engineers design steel-frame houses using the performance method and the prescriptive method of design.
3. Panelized construction is used to pre-build flat components such as walls, floors, and trusses.
4. Work gloves, ear protection, and safety glasses must be worn.

Section 26.1 On the Job

Answers will vary.

Section 26.2 Check Your Knowledge

1. Feathering is the process of attaching a screw to the screw gun bit without stopping the screw gun.
2. A drywall screw gun has a depth-sensitive nosepiece. The nosepiece prevents the bit from damaging the surface of the sheathing or wallboard while seating the screw.
3. The teeth of a steel cutting blade are carbide-tipped.
4. When layers of steel are screwed together, the first layer can "climb" the threads of the screw and pull away from the second layer. This is called "jacking."

Section 26.3 Check Your Knowledge

1. The pullout capacity of a screw is the ability of a screw to resist pulling out of a connection. Pullout capacity is based on the number of threads penetrating and holding the connection.
2. Self-tapping screws create their own holes and form their own threads as they engage.
3. Sheathing attached to walls must be held firmly against the steel before firing the pin or nail. The fasteners do not draw the sheathing against the steel as a screw does.
4. The steel and filler metals are melted together.

Section 26.3 On the Job

Answers will vary.

Chapter 26 Review Questions

1. Steel framing materials are treated with a hot-dipped galvanized coating to resist rust and corrosion.
2. The performance method depends on established engineering principles and design-load specifications. Architects and engineers use these principles and specifications to calculate the size and strength of individual steel framing members. The prescriptive method uses standardized tables that give specifications and other information. These tables are created using regional design codes, regional design-load data, structural limitation data, and knowledge of engineering practices. Load data for such things as seismic, snow, and wind loads are also determined by using tables for specific geographic regions.
3. Stick-built construction involves assembling individual studs, joists, and other members on site. Components for panelized construction are pre-built and then set in place on site.
4. The cut is very rough with sharp burrs.
5. A press brake is a tool used to create straight-line bends in steel.
6. Hand seamers are often called duck-billed pliers.
7. The hex washer head is the most common head style.
8. The two types of self-tapping tips used in steel framing screws are self-drilling and self-piercing.
9. An air gun is used.
10. Welding is the process of melting the steel and applying filler metals to fuse the pieces at the point of attachment. Welding is permanent. Clinching is the process of joining two layers of steel with pressure.

Instructional Plan: Framing Methods

Focus & Planning

Objectives
- Tell how to lay out steel floor joists.
- Describe the process for installing embedded or epoxied anchor bolts.
- Explain how to assemble a panelized wall.
- Explain how to set steel ceiling joists.

SAFETY FIRST
- ❏ Review the safety rules in the chapter.
- ❏ Make sure all necessary personal protective equipment is available and in good condition.
- ❏ Make sure all required tools and equipment are available and in good working order.

Instruction & Student Practice

Using the *Student Textbook*, have students:
- ❏ Read Section 27.1.
- ❏ Complete Section 27.1 Check Your Knowledge and On the Job.
- ❏ Read Section 27.2.
- ❏ Complete Section 27.2 Check Your Knowledge and On the Job.
- ❏ Read Section 27.3.
- ❏ Complete Section 27.3 Check Your Knowledge and On the Job.

Using *Carpentry Applications*, assign:
- ❏ 27-1: Framing a Wall with Steel
- ❏ 27-2: Framing a Box Header with Steel

Using *Carpentry Math*, assign:
- ❏ M27-1: Using the Construction Calculator in Roof Layout

Using the *Safety Guidebook*, assign:
- ❏ 2-1: Musculoskeletal Disorders

Using the *Instructor Productivity CD-ROM*, show PowerPoint presentation:
- ❏ 27-2: Framing a Box Header with Steel

Review & Student Performance

- ❏ Assign Review Questions from the end of the chapter.
- ❏ Assign the Chapter 27 Test or create a test using the *ExamView® Pro Test Generator* on the *Instructor Productivity CD-ROM*.
- ❏ Create and administer a Unit 5 Posttest using the *ExamView® Pro Test Generator* on the *Instructor Productivity CD-ROM*.
- ❏ Use *Carpentry Applications* job sheets as desired for final performance assessment.

(Continued on next page)

Answers to Textbook Questions

■ Section 27.1 Check Your Knowledge

1. A continuous joist spans the entire floor opening. A non-continuous joist is in two pieces. The pieces lap over an intermediate support.
2. The frame is attached using embedded or epoxied anchor bolts.
3. Bracing prevents the joists from rolling or twisting in the tracks.
4. Web stiffeners are pieces of stud or track material added to prevent joists from bending under the weight of floor loads.

■ Section 27.1 On the Job

Answers may vary but should note the fact that more material (studs and trusses as well as floor joists) will be required.

■ Section 27.2 Check Your Knowledge

1. An intersection is where two wall sections meet.
2. Temporary bracing should be installed every 8' to 12'.
3. In strong winds, workers can lose control and be injured.
4. Cripple studs are installed as necessary above rough openings.

■ Section 27.2 On the Job

Answers will vary.

■ Section 27.3 Check Your Knowledge

1. Minimal support bracing can provide more attic space. Fewer steel members are required, and costs are lower when roof designs are complicated.
2. The blocking keeps the joists from rolling in the tracks.
3. The common rafter method uses the length of the common rafters to set the ridge height. The calculation method uses the pitch of the roof and the rafter length to determine the ridge height.
4. He/she must check uplift wind-load safety data.

■ Section 27.3 On the Job

Answers will vary.

■ Chapter 27 Review Questions

1. Floor joists run in the same direction as the roof trusses. The layout may be from one end wall to another or from one side wall to another.
2. This enables both tracks to be marked for layout at the same time.
3. The flanges of the joists must all be oriented in the same direction, or on layout. The open side of the joist will be facing away from the starting point.
4. Embedded anchor bolts are set in place before a concrete foundation is poured. When the concrete cures and hardens, the concrete holds the bolt securely in place. Epoxied anchor bolts are installed in cured concrete or in concrete block. A hole is drilled into the foundation and filled with epoxy. A threaded rod is placed into the hole. When the epoxy hardens, it provides a strong bond that holds the bolt in place.
5. Clustering forms cold spots and reduces energy efficiency.
6. Find the door and window locations on the architectural drawings. Check the size of the openings. Mark the center of each opening on the top and bottom tracks. Add 12" to the width of the window openings. Using a tape measure, center the dimensions over the marks on the track. Mark the location at each end of the tape to indicate the location of the king studs. Place an X on the side of the mark away from the window. The webs of the king studs will be on the rough-opening side.
7. X-bracing is most effective.
8. Mark the layout of the top track. Start the layout on the same end for both sides. Measure and mark the layout over the headers. Also, mark the locations where there are no wall studs and the layout is not obvious. Install the ceiling joists in the tracks one at a time. Anchor the joists at the top of the track using two #10 screws. Install $2 \times 4 \times 33$ mil C-shape top flange bracing on the ceiling joists. Blocking must be installed every 12' OC.
9. The plumb cut at the top of the rafter must match the slope of the roof.
10. Wind uplift loads control the size of the rake extension.

Instructional Plan

Windows and Skylights

Focus & Planning

Objectives
- Describe the basic types of windows.
- Identify the ways in which windows are made energy efficient.
- Read a window schedule and a manufacturer's size table.
- Install a standard double-glazed or casement window.

SAFETY FIRST
- ❑ Review the safety rules in the chapter.
- ❑ Make sure all necessary personal protective equipment is available and in good condition.
- ❑ Make sure all required tools and equipment are available and in good working order.

Instruction & Student Practice

Using the *Student Textbook*, have students:
- ❑ Read Section 28.1.
- ❑ Complete Section 28.1 Check Your Knowledge and On the Job.
- ❑ Read Section 28.2.
- ❑ Complete Section 28.2 Check Your Knowledge and On the Job.

Using *Carpentry Applications*, assign:
- ❑ 28-1: Recognizing Window Types
- ❑ 28-2: Installing a Window

Using *Carpentry Math*, assign:
- ❑ M28-1: Ordering Windows and Doors

Using the *Safety Guidebook*, assign:
- ❑ 5-1: Hand Tools

Using the *Instructor Productivity CD-ROM*, show PowerPoint presentation:
- ❑ 28-1: Recognizing Window Types
- ❑ 28-2: Installing a Window

Review & Student Performance

- ❑ Assign Review Questions from the end of the chapter.
- ❑ Assign the Chapter 28 Test or create a test using the *ExamView® Pro Test Generator* on the *Instructor Productivity CD-ROM*.
- ❑ Create and administer a Unit 6 Pretest using the *ExamView® Pro Test Generator* on the *Instructor Productivity CD-ROM*.
- ❑ Use *Carpentry Applications* job sheets as desired for final performance assessment.

(Continued on next page)

Answers to Textbook Questions

■ **Section 28.1 Check Your Knowledge**

1. Every bedroom should have at least one window or an exterior door for an emergency escape.
2. The five basic types of windows are double-hung, casement, stationary, awning/hopper, and horizontal-sliding.
3. A hybrid window has a frame made from two or more materials.
4. The air trapped between the panes is what insulates the window.

■ **Section 28.1 On the Job**

Answers will vary.

■ **Section 28.2 Check Your Knowledge**

1. Rough opening dimensions are either located in the window schedule on the building plans or the window manufacturer's catalogs.
2. The flange is either part of the window unit or it is made of separate pieces that are inserted into grooves in the outer face of the jamb.
3. A mullion strip is a vertical member that separates the units in a combination window.
4. *Step flashing* uses small pieces of L-shaped metal interwoven with the roof shingles. *Pan flashing* consists of a one-piece metal assembly called a pan that fits over the skylight curb.

■ **Section 28.2 On the Job**

Answers will vary.

■ **Chapter 28 Review Questions**

1. It should be no less than 8 percent of the floor area.
2. *Any two*:
   - The sill height must be no higher than 44" above the floor.
   - The opening height must be no less than 24".
   - The opening width must be no less than 20".
   - The unobstructed opening must be no less than 5.7 sq. ft.
3. The five types include double-hung, casement, stationary, awning or hopper, and horizontal-sliding windows.
4. Both types of windows have a single sash that is hinged on one side and swings open.
5. Heat loss through metal is much greater than through similar wood units. Moisture-laden air inside the house can condense on the cold metal.
6. Low-emissivity (low-e) describes glass that radiates less heat to the outdoors than regular glass.
7. The air between the panes can be replaced with a material that insulates better, such as argon and krypton gases.
8. The glass size is the dimension of the glass, both visible and covered, in a single sash.
9. A *mullion* is a member that separates two adjacent units of a combination window. A *muntin* is a thin strip that holds and divides individual lights in a window sash.
10. The correct fastener is a 1 ¾" roofing nail.

Instructional Plan

Focus & Planning

Objectives
- Identify the various types of interior and exterior doors and door hardware.
- Handle a door properly at a job site.
- Identify the hand of any door.
- List at least three aspects of exterior door construction and installation that improve energy efficiency.
- Install an interior pre-hung door.

SAFETY FIRST
- ❑ Review the safety rules in the chapter.
- ❑ Make sure all necessary personal protective equipment is available and in good condition.
- ❑ Make sure all required tools and equipment are available and in good working order.

Instruction & Student Practice

Using the *Student Textbook*, have students:
- ❑ Read Section 29.1.
- ❑ Complete Section 29.1 Check Your Knowledge and On the Job.
- ❑ Read Section 29.2.
- ❑ Complete Section 29.2 Check Your Knowledge and On the Job.
- ❑ Read Section 29.3.
- ❑ Complete Section 29.3 Check Your Knowledge and On the Job.

Using *Carpentry Applications*, assign:
- ❑ 29-1: Installing a Pre-Hung Interior Door
- ❑ 29-2: Installing a Lockset

Using *Carpentry Math*, assign:
- ❑ M29-1: Setting Up Proportions
- ❑ M29-2: Using Proportions in Estimating Materials
- ❑ M29-3: Solving Inverse Proportions
- ❑ M29-4: Using Proportions to Estimate Labor Costs

Using the *Safety Guidebook*, assign:
- ❑ 2-1: Musculoskeletal Disorders
- ❑ 5-13: Circular Saws

Using the *Instructor Productivity CD-ROM*, show PowerPoint presentation:
- ❑ 29-1: Setting a Pre-Hung Interior Door
- ❑ 29-2: Installing a Lockset

Review & Student Performance

- ❑ Assign Review Questions from the end of the chapter.
- ❑ Assign the Chapter 29 Test or create a test using the *ExamView® Pro Test Generator* on the *Instructor Productivity CD-ROM*.
- ❑ Use *Carpentry Applications* job sheets as desired for final performance assessment.

(Continued on next page)

Answers to Textbook Questions

Section 29.1 Check Your Knowledge
1. The most common type of door is a flat-panel or raised-panel passage door.
2. Hollow-core construction consists of a light framework faced with thin plywood or hardboard.
3. It is 4 9/16" wide.
4. The hinge most frequently used for hanging residential doors is the loose-pin butt mortise hinge.

Section 29.1 On the Job
Answers will vary.

Section 29.2 Check Your Knowledge
1. The opening should be approximately 3" wider and 2" higher than the door.
2. The hinge-butt gauge is used to lay out the position of a hinge leaf. When it is removed, a chisel is used to cut the hinge gains. A butt-hinge template is also used to lay out the position of a hinge leaf, but it guides a router that actually cuts the gains.
3. The bottom edge of the lowest hinge should be 11" from the bottom of the door. The top edge of the upper hinge should be 7" from the top of the door.
4. Two holes should be bored. A large hole should be bored in the face of the door and a smaller one in the edge of the door.

Section 29.2 On the Job
Answers will vary.

Section 29.3 Check Your Knowledge
1. Most interior passage doors are 1 3/8" thick.
2. A pocket door is a door that slides into a "pocket" inside a wall.
3. It ensures that the side jambs will be a consistent distance apart.
4. To prevent the veneer from splintering when the door is trimmed, some carpenters use a cutting guide or a blade specifically designed for crosscutting plywood.

Section 29.3 On the Job
Answers will vary.

Chapter 29 Review Questions
1. *Exterior, any 3:* Sliding-glass, French, glazed, fire-rated, and garage. *Interior, any 3:* Louvered, sliding, bifold, folding, and pocket.
2. The basic parts of panel doors are stiles (vertical wood members), rails (horizontal wood members), and panels (members filling spaces between the stiles and the rails).
3. *Any five of the following*:
   - Doors should not be delivered to the house until "wet" materials, such as plaster and concrete, have given up most of their moisture.
   - Keep all doors away from unusual heat or dryness. Sudden changes, such as heat forced into a building to dry it out, should be avoided.
   - Store doors under cover in a clean, dry, well-ventilated area.
   - Handle doors with clean gloves; bare hands leave finger marks and soil stains on surfaces that have not yet been sealed or painted. When moving a door, carry it. Do not drag it.
   - Store doors on edge on a level surface.
   - Seal the top and bottom edges of wood doors immediately to prevent moisture from reaching end grain.
   - Condition wood doors to the average local moisture content before hanging.
   - Apply a finish to the doors as soon as possible after they have been installed.
4. From the side on which the hinges are not visible when the door is closed, stand with your back to the hinge jamb. A door that swings to your right is a right-hand door; one that swings to your left is a left-hand door.
5. Building codes require this type between a house and an attached garage.
6. It should have a core of rigid insulation and very effective weatherstripping. It should also be fitted with a storm door.
7. The clearance should be 1/16" at the sides and on top.
8. The standard height is 6'-8".
9. They are made from wood, reinforced vinyl, or plastic-coated wood.
10. Clearances for a pre-hung interior door are top 1/8", bottom 1/2" or more, hinge side 1/16", latch side 1/8".

Instructional Plan

---- Focus & Planning ----

Objectives
- Recognize different roofing products.
- List the basic steps for installing a strip shingle roof covering.
- Describe the various types of flashing and explain where they are used.
- Tell the difference between shakes and wood shingles.

SAFETY FIRST
- ❏ Review the safety rules in the chapter.
- ❏ Make sure all necessary personal protective equipment is available and in good condition.
- ❏ Make sure all required tools and equipment are available and in good working order.

---- Instruction & Student Practice ----

Using the *Student Textbook*, have students:
- ❏ Read Section 30.1.
- ❏ Complete Section 30.1 Check Your Knowledge and On the Job.
- ❏ Read Section 30.2.
- ❏ Complete Section 30.2 Check Your Knowledge and On the Job.
- ❏ Read Section 30.3.
- ❏ Read Estimating Roofing Materials and complete Section 30.3 Estimating On the Job.
- ❏ Complete Section 30.3 Check Your Knowledge and On the Job.
- ❏ Read Section 30.4.
- ❏ Complete Section 30.4 Check Your Knowledge and On the Job.

Using *Carpentry Applications*, assign:
- ❏ 30-1: Figuring the Number of Roof Shingles Using a Construction Calculator
- ❏ 30-2: Laying Roof Shingles

Using *Carpentry Math*, assign:
- ❏ Estimating Waterproof Roofing Materials

Using the *Safety Guidebook*, assign:
- ❏ 4-1: Ladders
- ❏ 4-3: Working on Roofs
- ❏ 5-25: Power Nailers and Staplers

Using the *Instructor Productivity CD-ROM*, show PowerPoint presentation:
- ❏ 30-2: Laying Shingles

---- Review & Student Performance ----

- ❏ Assign Review Questions from the end of the chapter.
- ❏ Assign the Chapter 30 Test or create a test using the *ExamView® Pro Test Generator* on the *Instructor Productivity CD-ROM*.
- ❏ Use *Carpentry Applications* job sheets as desired for final performance assessment.

(Continued on next page)

Answers to Textbook Questions

■ Section 30.1 Check Your Knowledge
1. The amount of roofing is a square.
2. Coverage indicates the amount of weather protection provided by the overlapping of shingles, based on how many layers of material there are. Exposure is the amount of the shingle that shows after installation.
3. This speeds the installation and reduces strain on the roofer's arm and hand.
4. Many workers, particularly roofers, are injured each year in falls.

■ Section 30.1 On the Job
Answers will vary.

■ Section 30.2 Check Your Knowledge
1. It protects the sheathing from moisture until the shingles can be applied. It provides a second layer of weather protection. It prevents direct contact between asphalt shingles and wood resin from the sheathing, which may damage the shingles.
2. Eaves protection should extend from the end of the eaves (the edge of the roof) to a point that is at least 2' inside the exterior wall line of the house.
3. The rib reduces the tendency for water to pour down one side of the roof and splash up on the adjacent side.
4. Step flashing consists of individual L-shaped pieces of metal. It is used where a vertical surface, such as a chimney or wall, meets the roof.

■ Section 30.2 On the Job
Answers will vary.

■ Section 30.3 Estimating on the Job
29 squares

■ Section 30.3 Check Your Knowledge
1. This is a process in which bundles of shingles are carried to the roof and distributed evenly.
2. A row of inverted shingles or rolled roofing can be used for the starter course.
3. The typical exposure is 5".
4. A closed valley is a valley in which strip shingles are interwoven to protect the valley from seepage. No metal flashing is necessary.

■ Section 30.4 Check Your Knowledge
1. Spaced, or open, sheathing is most commonly used beneath wood shingles and shakes.
2. For starter courses, #3 grade and undercourse grade are used.
3. Wood shingles should be spaced at least ¼" apart to provide for expansion.
4. Nails at least two sizes larger than the nails used to apply the shingles are required.

■ Section 30.4 On the Job
Answers will vary.

■ Chapter 30 Review Questions
1. It has a roof slope of 4-in-12 or greater.
2. Roofing suitable for large areas includes roll roofing, built-up roofing, metal roofing, and single-ply roofing.
3. Interlocking (lock-down) shingles, or standard three-tab shingles attached with six nails instead of the standard four nails are suitable for high-wind areas.
4. To minimize damage caused by ice dams, provide eaves flashing.
5. Flashing is required wherever the roof covering intersects another surface, such as a wall, chimney, skylight, or vent pipe. It is sometimes also used in the valleys.
6. This is corrosion caused by electrolysis. Tiny electrical currents are created when different metals are in contact with each other and with water.
7. Nails should be made of hot-dipped galvanized steel, aluminum, or stainless steel.
8. Three-tab shingles require four nails for each strip. When the shingles are applied with a 5" exposure, the four nails are placed ⅝" above the top of the cutouts. The nails are located horizontally with one nail 1" back from each end, and one nail on the centerline of each cutout.
9. Wood shingles are thinner than wood shakes.
10. Solid sheathing should be applied above the eave line up to a point at least 24" inside the interior wall line of the building. The solid sheathing should then be covered either with a double layer of No. 15 asphalt-saturated felt or with a comparable product, such as self-adhering bitumen sheets.

Instructional Plan

CHAPTER 31: Roof Edge Details

---Focus & Planning---

Objectives
- Identify different types of cornice construction.
- Assemble a simple box cornice.
- Explain the purpose of a cornice return.
- Identify the main parts of a gutter system.

SAFETY FIRST
- ❑ Review the safety rules in the chapter.
- ❑ Make sure all necessary personal protective equipment is available and in good condition.
- ❑ Make sure all required tools and equipment are available and in good working order.

---Instruction & Student Practice---

Using the *Student Textbook*, have students:
- ❑ Read Section 31.1.
- ❑ Complete Section 31.1 Check Your Knowledge and On the Job.
- ❑ Read Section 31.2.
- ❑ Complete Section 31.2 Check Your Knowledge and On the Job.
- ❑ Read Section 31.3.
- ❑ Complete Section 31.3 Check Your Knowledge and On the Job.

Using *Carpentry Applications*, assign:
- ❑ 31-1: Installing Lookouts and Fascia
- ❑ 31-2: Installing a Soffit

Using *Carpentry Math*, assign:
- ❑ M31-1: Estimating J-Channel
- ❑ M31-2: Estimating Undersill Trim
- ❑ M31-3: Estimating Corner Pieces
- ❑ M31-4: Estimating F-Channel

Using the *Safety Guidebook*, assign:
- ❑ 4-1: Ladders
- ❑ 4-2: Scaffolds
- ❑ 4-3: Working on Roofs

---Review & Student Performance---

- ❑ Assign Review Questions from the end of the chapter.
- ❑ Assign the Chapter 31 Test or create a test using the *ExamView® Pro Test Generator* on the *Instructor Productivity CD-ROM*.
- ❑ Use *Carpentry Applications* job sheets as desired for final performance assessment.

(Continued on next page)

Answers to Textbook Questions

■ Section 31.1 Check Your Knowledge

1. The three basic types are open, box, and closed.
2. The underside of the roof sheathing is exposed, and any imperfections may be unattractive. To counter this problem, carpenters must install a higher grade of plywood sheathing where it will show.
3. Before adding a box cornice, a carpenter must check the plumb cuts on the rafter tails to make certain they are all in line with one another.
4. These materials require little maintenance, are entirely prefinished, are light in weight, and easy to install.

■ Section 31.1 On the Job

Frieze: 110 lf + 10% = 121 lf. Round to 122 lf.
Fascia: 122 lf + 10% = 134.2 lf. Round to 136 lf.

■ Section 31.2 Check Your Knowledge

1. The fly rafter helps to support an extended rake.
2. It is usually made with a header rafter on the inside and a fly rafter on the outside. Each is nailed to the ends of lookouts. The header rafter is face-nailed directly to the standard rafters with pairs of 12d nails spaced 16" to 20" apart. Each lookout should be toenailed to the rake wall plate.
3. The cornice return is a decorative detail that provides a transition between the cornice and the rake.
4. It is a curved piece of wood that can be attached to the underside of the rake trim on houses with an open cornice.

■ Section 31.2 On the Job

Answers will vary.

■ Section 31.3 Check Your Knowledge

1. The two types are the formed-metal gutter and the half-round gutter.
2. Downspouts are usually corrugated for added strength.
3. The gutters should slope at least 1" every 16' toward the downspouts.
4. A ferrule is a short metal tube that is placed between the inner and outer faces of the gutter. Long aluminum spikes are then driven through the face of the gutter, through the ferrule, and into the fascia.

■ Section 31.3 On the Job

Answers will vary.

■ Chapter 31 Review Questions

1. With a box cornice, the rafter tails are entirely enclosed by the roof sheathing, fascia, and soffit material. With an open cornice the rafter tails are exposed.
2. The materials are more uniform than solid lumber, free of defects, and consistent in quality. They are available in long lengths. Many are pre-primed at the factory on all four surfaces, which saves labor at the job site and improves durability.
3. If the soffit is narrow, the connection between the two may be a butt joint. If the soffit is wide, carpenters often fit one edge of the soffit material into a groove cut in the back of the fascia.
4. A pork chop is a curved piece of wood attached to the underside of the rake trim to form a decorative detail on a house with open cornices.
5. Lookouts are usually spaced 16" or 24" OC.
6. A cornice return provides a transition between the rake and a cornice on a house with a gable roof.
7. The main parts include the gutters, downspouts or leaders, and splash blocks.
8. The two most common materials in the construction of gutters are aluminum and copper.
9. Gutters are fabricated on site, using machines that form them in almost any length from continuous coils of flat aluminum stock.
10. At least two straps should be used to secure an 8' length of downspout.

Instructional Plan

CHAPTER 32: Siding

Focus & Planning

Objectives
- Prevent moisture from seeping into or behind siding.
- Install plain-bevel wood siding.
- Describe the four coursing styles for wood shingles.
- Use good nailing technique on vinyl siding.

SAFETY FIRST
- ❑ Review the safety rules in the chapter.
- ❑ Make sure all necessary personal protective equipment is available and in good condition.
- ❑ Make sure all required tools and equipment are available and in good working order.

Instruction & Student Practice

Using the *Student Textbook*, have students:
- ❑ Read Section 32.1.
- ❑ Complete Section 32.1 Check Your Knowledge and On the Job.
- ❑ Read Section 32.2.
- ❑ Read Estimating Beveled Siding Materials and complete Section 32.2 Estimating on the Job.
- ❑ Complete Section 32.2 Check Your Knowledge and On the Job.
- ❑ Read Section 32.3.
- ❑ Complete Section 32.3 Check Your Knowledge and On the Job.
- ❑ Read Section 32.4.
- ❑ Complete Section 32.4 Check Your Knowledge and On the Job.

Using *Carpentry Applications*, assign:
- ❑ 32-1: Estimating Siding
- ❑ 32-2: Applying Vinyl Siding

Using *Carpentry Math*, assign:
- ❑ M32-1: Estimating Siding Start Strip
- ❑ M32-2: Estimating Siding
- ❑ M32-3: Estimating Soffit
- ❑ M32-4: Estimating Fascia

Using the *Safety Guidebook*, assign:
- ❑ 5-15: Radial-Arm Saws
- ❑ 5-16: Miter Saws
- ❑ 6-3: Hazardous Materials

Using the *Instructor Productivity CD-ROM*, show PowerPoint presentation:
- ❑ 32-1: Estimating Siding
- ❑ 32-2: Applying Vinyl Siding

Review & Student Performance

- ❑ Assign Review Questions from the end of the chapter.
- ❑ Assign the Chapter 32 Test or create a test using the *ExamView® Pro Test Generator* on the *Instructor Productivity CD-ROM*.
- ❑ Use *Carpentry Applications* job sheets as desired for final performance assessment.

(Continued on next page)

Answers to Textbook Questions

Section 32.1 Check Your Knowledge

1. Woods used for siding should have the ability to accept paint or stains, be easy to work with, and be dimensionally stable.
2. The cedars, eastern white pine, western white pine, sugar pine, cypress, and redwood have a high degree of the properties.
3. The horizontal joints between courses of beveled siding should never be caulked. Caulking prevents moisture vapor from escaping from behind the siding, which can cause rot.
4. They protect the sheathing from wind-blown rain that may get behind siding or trim.

Section 32.1 On the Job
Answers will vary.

Section 32.2 Estimating on the Job
1,800 sq. ft. of siding and 13.5 lbs. of nails

Section 32.2 Check Your Knowledge

1. This is giving a coat of primer to the back side of siding prior to application. It improves the durability of the siding and the finish.
2. Boards may be treated with a water repellent. The ends of siding boards cut during installation should also receive a liberal treatment with a water repellent.
3. A butt joint or a scarf joint is used.
4. A small wood block is used for accurately marking siding pieces that must fit against a vertical surface.

Section 32.2 On the Job
Thirteen courses are required.

Section 32.3 Check Your Knowledge

1. The first grade includes clear, all heartwood shingles. The second grade consists of shingles with a clear exposed area and allows defects in that part normally covered in use.
2. The undercourse is a lower-grade layer of shingles that will not be exposed to the weather.
3. Because the undercourse is partially exposed, it has to be the same grade as the outer course.
4. They come in lengths of 16", 18", and 24".

Section 32.3 On the Job
Answers will vary.

Section 32.4 Check Your Knowledge

1. Vinyl siding expands and contracts much more than wood.
2. Leave approximately $\frac{1}{32}$" (the thickness of a dime) between the underside of the nail head and the vinyl.
3. Plywood panel siding serves as both sheathing and exterior wall covering.
4. Cutting, drilling, or sanding fiber cement releases a cloud of fine dust that contains silica. Use of a dust mask or NIOSH-approved respirator is recommended.

Section 32.4 On the Job
Answers will vary.

Chapter 32 Review Questions

1. The use of proper construction detailing, suitable flashing, and high-quality caulking prevents moisture from seeping into or behind any siding.
2. Vapor-resistant materials might trap moisture behind the siding or against the sheathing.
3. Exposure is the amount of surface of each board that is exposed to the weather and is based on the board's width.
4. It speeds the process and ensures a uniform layout from wall to wall around the house.
5. They cut the end board of each course (the closure board) approximately $\frac{1}{16}$" longer than the actual distance. Then they bow the piece slightly to get the ends in position and push the middle into place.
6. The four styles are single coursing, double coursing, ribbon coursing, and decorative coursing.
7. Nails should be placed in the center of slots in the panel's nailing flange.
8. It is used around windows, doors, and other locations to receive the ends of vinyl siding panels.
9. The edges and ends of plywood siding panels should be sealed with primer or a water-repellent sealant prior to installation.
10. Metal lath holds stucco in place.

Instructional Plan

Chapter 33: Brick-Veneer Siding

Focus & Planning

Objectives
- Identify the tools used in working with brick and mixing mortar.
- Cut brick with a mason's hammer.
- Name the type of mortar used most often for brick veneer.
- Explain why care should be taken when laying brick in cold weather.

SAFETY FIRST
- ☐ Review the safety rules in the chapter.
- ☐ Make sure all necessary personal protective equipment is available and in good condition.
- ☐ Make sure all required tools and equipment are available and in good working order.

Instruction & Student Practice

Using the *Student Textbook*, have students:
- ☐ Read Section 33.1.
- ☐ Complete Section 33.1 Check Your Knowledge and On the Job.
- ☐ Read Section 33.2.
- ☐ Complete Section 33.2 Check Your Knowledge and On the Job.

Using *Carpentry Applications*, assign:
- ☐ 33-1: Estimating Brick
- ☐ 33-2: Building a Brick-Veneer Wall

Using *Carpentry Math*, assign:
- ☐ M33-1: Estimating Concrete
- ☐ M33-2: Estimating Brick

Using the *Safety Guidebook*, assign:
- ☐ 2-2: Personal Protective Equipment
- ☐ 5-1: Hand Tools
- ☐ 5-21: Jointers
- ☐ 6-3: Hazardous Materials
- ☐ 6-4: Bulk Lifting/Moving Operations

Review & Student Performance

- ☐ Assign Review Questions from the end of the chapter.
- ☐ Assign the Chapter 33 Test or create a test using the *ExamView® Pro Test Generator* on the *Instructor Productivity CD-ROM*.
- ☐ Create and administer a Unit 6 Posttest using the *ExamView® Pro Test Generator* on the *Instructor Productivity CD-ROM*.
- ☐ Use *Carpentry Applications* job sheets as desired for final performance assessment.

(Continued on next page)

Answers to Textbook Questions

■ Section 33.1 Check Your Knowledge

1. The heel is the widest portion.
2. Wear suitable eye protection when cutting brick. If cutting with a masonry saw, wear a dust mask so as not to breathe in the fine brick dust.
3. Maintenance required for a mortar mixer includes cleaning the mixing drum thoroughly after use, checking any drive belts for proper tension, and ensuring that the retractable guard over the mixing drum works smoothly. Mortar should not be allowed to dry on the mixer's moving parts.
4. The dry ingredients must be measured and then blended.

■ Section 33.1 On the Job

Answers will vary.

■ Section 33.2 Check Your Knowledge

1. A brick veneer wall must be supported by a masonry or concrete foundation. A supporting 5" wide ledge or shelf is made in the main-house foundation.
2. The wire-type tie is more corrosion resistant.
3. A lead corner is a partially constructed corner of a brick-veneer wall.
4. Brick is laid to a line in order to ensure that each course of brick is level and straight.

■ Section 33.2 On the Job

Efflorescence is caused when water-soluble salts are brought to the surface of the masonry by rain water. When the water evaporates, the salts are left on the surface of the brick. The salts may be present in the brick itself, in the mortar, or in materials behind the brickwork.

Efflorescence can be prevented by using low-alkaline mortar, protecting partially-completed walls from rain, and by storing brick off the ground and under cover. This avoids contamination by dirt or ground water.

Removing efflorescence is relatively easy because most efflorescent salts are water-soluble. They can be removed by scrubbing with a stiff bristle brush (not a wire brush) and clear water. Heavy accumulations can be removed with chemical cleaners specifically made for the purpose.

■ Chapter 33 Review Questions

1. The standard white mason's rule is used for measuring standard or modular brick. An oversized yellow mason's rule is easier to use when oversized bricks are measured.
2. Brick is sometimes cut with a masonry saw or a brick hammer. If a more accurate cut must be made, a brick hammer and a brick set may be used.
3. Building brick, facing brick, and fire brick are the three basic types of brick.
4. When specifying the size of a brick, the dimensions should always be listed in the following order: thickness by height by length.
5. A modular brick includes an allowance for the thickness of a standard mortar joint.
6. Type N is a general-purpose mortar for brick-veneer walls.
7. The mason's hoe has holes in the blade to make mixing mortar easier.
8. The top edge of the flashing should be slipped behind the building paper or building felt already attached to the sheathed walls.
9. Weep holes provide drainage. They are located near the bottom of the wall.
10. Cold weather slows the hydration process in mortar and may affect the strength of the wall. Mortar that freezes is less weather resistant. Freezing can even reduce or destroy the bond between brick and mortar.

Instructional Plan

Stairways

Focus & Planning

Objectives
- Identify the method of construction used on any stairway.
- Understand the building code requirements that apply to stairs.
- Tell the purpose of the different parts of a stairway.
- Tell how to lay out a cut-stringer stairway.
- Tell how to lay out a cleat-stringer stairway.

SAFETY FIRST
- ❏ Review the safety rules in the chapter.
- ❏ Make sure all necessary personal protective equipment is available and in good condition.
- ❏ Make sure all required tools and equipment are available and in good working order.

Instruction & Student Practice

Using the *Student Textbook*, have students:
- ❏ Read Section 34.1.
- ❏ Complete Section 34.1 Check Your Knowledge and On the Job.
- ❏ Read Section 34.2.
- ❏ Complete Section 34.2 Check Your Knowledge and On the Job.
- ❏ Read Section 34.3.
- ❏ Read Estimating Stairway Materials and Lumber and complete Section 34.3 Estimating on the Job.
- ❏ Complete Section 34.3 Check Your Knowledge and On the Job.

Using *Carpentry Applications*, assign:
- ❏ 34-1: Figuring a Stair Stringer Using a Construction Calculator
- ❏ 34-2: Laying Out a Stairway Stringer

> **Carpenter's Tip**
>
> Use 10" rolls of kraft or other inexpensive paper to practice the layout of the stairway.

Using *Carpentry Math*, assign:
- ❏ M34-1: Laying out a Stairway
- ❏ M34-2: Estimating Stairway Stringer Length
- ❏ M34-3: Estimating Stairway Labor Costs
- ❏ M34-4: Using the Construction Calculator in a Stair Layout

Using the *Instructor Productivity CD-ROM*, show PowerPoint presentation:
- ❏ 34-2: Laying Out a Stairway Stringer

Review & Student Performance

- ❏ Assign Review Questions from the end of the chapter.
- ❏ Assign the Chapter 34 Test or create a test using the *ExamView® Pro Test Generator* on the *Instructor Productivity CD-ROM*.
- ❏ Create and administer a Unit 7 Pretest using the *ExamView® Pro Test Generator* on the *Instructor Productivity CD-ROM*.
- ❏ Use *Carpentry Applications* job sheets as desired for final performance assessment.

(Continued on next page)

Answers to Textbook Questions

■ Section 34.1 Check Your Knowledge

1. A stairwell is the vertical shaft containing a stairway.
2. A riser is the vertical portion of a step.
3. The two types are cleat- and cut-stringer stairs. A cleat-stringer stair has a pair of stringers and a series of plank treads supported by cleats attached to the sides of each stringer. Cut-stringer stairs have treads and risers that fit into notches sawn into the upper edge of the stringers. Stairs of this type usually have two or three stringers, but could have more.
4. Total rise is the vertical distance from the surface of one floor to the surface of the next floor.

■ Section 34.1 On the Job

Answers will vary.

■ Section 34.2 Check Your Knowledge

1. The minimum headroom required by code is 6'-8".
2. The minimum width for a main stairway is measured between the finished walls of the stairwell.
3. If the risers are too high, climbing the steps can be tiring. If the treads are too shallow, toes will bump the riser at each step.
4. A handrail should be between 34" and 38" high. The height is measured vertically from the upper edge of the nosing to the top of the handrail.

■ Section 34.2 On the Job

Answers will vary.

■ Section 34.3 Estimating on the Job

15.25 hours, $343.00

■ Section 34.3 Check Your Knowledge

1. The total rise is the vertical distance between the surface of the lower finish floor and the surface of the upper finish floor.
2. The total number of treads depends on the manner in which the upper end of the stairway is anchored to the upper landing.
3. Do not overcut the stringer when making intersecting cuts, because this weakens it. Instead, finish the cuts with a handsaw or jigsaw.
4. The handrail is attached to the wall with adjustable metal brackets screwed to the stairwell framing.

On the Job

Answers will vary.

■ Chapter 34 Review Questions

1. Building codes regulate dimensions and design because so many people are injured in stairway accidents.
2. You would identify a cut-stringer stair by the notches cut in the stringers.
3. Open (plain) stringers are cut to follow the lines of the treads and risers. Closed stringers have risers and treads mortised into them.
4. A stairway is turned to conserve space. Stairways that turn usually incorporate a landing or radiating treads called winders.
5. Headroom is the clearance above a step. It is measured from the outside edge of the nosing to the surface of the ceiling directly overhead.
6. Unit run is the distance from the face of one riser to the face of the next, or the width of a tread less the nosing.
7. The building code allows a maximum riser height of 7 ¾".
8. A handrail can prevent a stair user from tumbling down an entire flight of stairs, thus avoiding serious injury.
9. A third stringer should be installed in the middle of the stairs when the treads are less than 1 ⅛" thick or if the stairs are more than 2'-6" wide.
10. The skirtboard is a finished board that is nailed to the wall before the stringers are installed. It provides a finished edge against the wall, and makes it easier to paint or wallpaper the adjacent plaster or drywall.

Instructional Plan

Chapter 35: Molding and Trim

— Focus & Planning —

Objectives
- Identify uses for molding and trim other than decoration.
- Tell which joints are used for moldings and trims and why.
- Identify different types of molding and trim.
- Tell how to scribe molding and trim to an uneven surface.
- Explain how to cut a coped joint.

SAFETY FIRST
- ❏ Review the safety rules in the chapter.
- ❏ Make sure all necessary personal protective equipment is available and in good condition.
- ❏ Make sure all required tools and equipment are available and in good working order.

— Instruction & Student Practice —

Using the *Student Textbook*, have students:
- ❏ Read Section 35.1.
- ❏ Complete Section 35.1 Check Your Knowledge and On the Job.
- ❏ Read Section 35.2.
- ❏ Complete Section 35.2 Check Your Knowledge and On the Job.
- ❏ Read Section 35.3.
- ❏ Read Estimating Molding and Trim and Complete Section 35.3 Estimating on the Job.
- ❏ Complete Section 35.3 Check Your Knowledge and On the Job.

Using *Carpentry Applications*, assign:
- ❏ 35-1: Installing Base Moldings
- ❏ 35-2: Installing Crown Molding

Using the *Safety Guidebook*, assign:
- ❏ 2-1: Musculoskeletal Disorders

Carpenter's Tip
- When first driving nails into hardwood trim, you can help avoid splitting the wood if you first blunt the nail with a hammer.
- A piece of ¼" thick underlayment makes a good "gauge" for a reveal.
- There are special tools for cutting crown molding. One is the Delta saw buck.

Using the *Instructor Productivity CD-ROM*, show PowerPoint presentation:
- ❏ 35-1: Installing Base Molding
- ❏ 35-2: Installing Crown Molding

— Review & Student Performance —

- ❏ Assign Review Questions from the end of the chapter.
- ❏ Assign the Chapter 35 Test or create a test using the *ExamView® Pro Test Generator* on the *Instructor Productivity CD-ROM*.
- ❏ Use *Carpentry Applications* job sheets as desired for final performance assessment.

(Continued on next page)

Answers to Textbook Questions

■ Section 35.1 Check Your Knowledge

1. Trim carpentry is all the woodwork installed in a building.
2. The abbreviation S4S stands for "surfaced four sides."
3. It should be smooth, close-grained, and free from pitch streaks.
4. The recommended moisture content is 8 percent.

■ Section 35.1 On the Job

Answers will vary.

■ Section 35.2 Check Your Knowledge

1. The two are the miter cut and the square cut.
2. A reveal is a small offset between a piece of trim and the surface it is applied to. The small step this creates adds visual interest. It also allows the trim carpenter to make adjustments to the fit of the casing if the door or window is not perfectly square.
3. The window is surrounded by four lengths of casing that are mitered together.
4. Cut the apron to a length that matches the distance between the outer edges of the side casings.

■ Section 35.2 On the Job

Answers will vary.

■ Section 35.3 Estimating on the Job

Crown = 59 l.f. + 10% (5.9 l.f.) = 64.9 l.f., rounded to 66 l.f. Baseboard = 56 l.f. + 10% (5.6 l.f.) = 61.6 l.f., rounded to 62.

■ Section 35.3 Check Your Knowledge

1. The shoe should be nailed into the baseboard, not into the flooring.
2. Two methods are nailing across the joint and securing it with a biscuit that has been glued into position.
3. Baseboard should be scribed to a wall when the wall is out of plumb.
4. Crown molding is a fairly large sprung molding that usually includes both curved and angular surfaces. It calls for special cutting and installation techniques.

■ Section 35.3 On the Job

Answers will vary.

■ Chapter 35 Review Questions

1. Traditionally, molding has referred to narrow lengths of wood with a shaped or curved profile, while trim has referred more to straight lengths of wood, such as a 1×4, surfaced on four sides. However, the two terms are often used interchangeably. Trim may also be used as a verb.
2. Molding and trim are reinforcements for framing, and they protect walls, reinforce shelving, and cover gaps.
3. Casing is all the trim around a door or window. The stool laps a window sill and extends beyond the casing. An apron is a finish member below the stool.
4. The three are miter joints, coped joints, and butt joints. A miter joint is used where square-edged baseboards meet at inside corners, and for all baseboards that meet at outside corners. A coped joint is used where two lengths of profiled molding intersect at an inside corner. A butt joint is used where any baseboard meets the wall at an inside corner.
5. When carpeting is to be installed, the baseboard is installed first, using temporary spacers to lift it slightly above the subfloor. The edges of the carpet are then tucked beneath the baseboard.
6. When it is necessary to use more than one length of molding along a wall, join the pieces over a wall stud, using a mitered lap joint.
7. You should do the following:
   - Set a piece of baseboard against the wall. Mark a layout line on the floor along the edge of the piece.
   - Repeat the process on the adjoining wall.
   - Hold the first piece to be mitered in place. Mark it where it intersects the layout line.
   - Set the miter saw to a 45° bevel angle and cut just outside the layout line.
   - Hold the second piece in place and mark it where it intersects the layout line. Cut the piece at a 45° bevel angle.
   - Test fit the pieces and trim them as needed.
8. The first piece is installed. The end of the second piece is placed against its face. A line is scribed parallel to the face of the first piece. The scribed piece is cut along the line.
9. The angle at which sprung molding projects away from the wall is called the springing angle.
10. Miter the end at 45°. Hold the coping saw at 90° to the back of the molding and cut along the edge left by the miter cut.

Instructional Plan

CHAPTER 36: Cabinets and Countertops

Focus & Planning

Objectives
- Identify the five basic kitchen layouts.
- Explain the difference between frameless and face-frame cabinet construction.
- Explain how to install a base cabinet.
- Explain how to install a wall cabinet.
- Apply plastic laminate to a countertop surface.

SAFETY FIRST
- ❏ Review the safety rules in the chapter.
- ❏ Make sure all necessary personal protective equipment is available and in good condition.
- ❏ Make sure all required tools and equipment are available and in good working order.

Instruction & Student Practice

Using the *Student Textbook*, have students:
- ❏ Read Section 36.1.
- ❏ Complete Section 36.1 Check Your Knowledge and On the Job.
- ❏ Read Section 36.2.
- ❏ Complete Section 36.2 Check Your Knowledge and On the Job.
- ❏ Read Section 36.3.
- ❏ Complete Section 36.3 Check Your Knowledge and On the Job.
- ❏ Read Estimating Cabinetry and complete Section 36.3 Estimating on the Job.

Using *Carpentry Applications*, assign:
- ❏ 36-1: Building Workshop Cabinets
- ❏ 36-2: Applying Plastic Laminate to a Countertop

> **Carpenter's Tip**
>
> When planning a kitchen, it's a good idea to specify engineered lumber for the wall studs. If using regular lumber, the carpenter should turn all the studs so that the crowns face in the same direction. Doing so will greatly help with the installation of cabinets.

Using *Carpentry Math*, assign:
- ❏ M36-1: Estimating Countertop Coverings

Using the *Safety Guidebook*, assign:
- ❏ 5-19: Routers
- ❏ 5-20: Sanders
- ❏ 6-3: Hazardous Materials

Review & Student Performance

- ❏ Assign Review Questions from the end of the chapter.
- ❏ Assign the Chapter 36 Test or create a test using the *ExamView® Pro Test Generator* on the *Instructor Productivity CD-ROM*.
- ❏ Use *Carpentry Applications* job sheets as desired for final performance assessment.

(Continued on next page)

Answers to Textbook Questions

■ Section 36.1 Check Your Knowledge

1. The basic arrangement of cabinets in any room, including the arrangement of the appliances and related plumbing fixtures, can be found on building plans.
2. This is a design philosophy aimed at making a house usable and safe for the widest variety of people, including the elderly and those with disabilities.
3. There are five basic kitchen layouts:
   - The U-shaped kitchen has the sink at the bottom of the U and the range and refrigerator on opposite sides.
   - The L-shaped kitchen has the sink and range on one leg and the refrigerator on the other. Sometimes the dining space is located in the opposite corner.
   - The parallel-wall kitchen is often used where there is limited space. Appliances are on opposite walls.
   - In the side-wall kitchen the cabinets, sink, range, and refrigerator are located on one wall.
   - The island kitchen features a cabinet "island" that is separate from the main cabinet runs.
4. A kitchen wall cabinet is usually 12" deep.

■ Section 36.1 On the Job
Answers will vary.

■ Section 36.2 Check Your Knowledge

1. Stock cabinets are built in standard sizes and in width increments of 3". Semi-custom cabinets are built only when they are ordered for a specific house. A client has more choices about style, finishes, and hardware. Semi-custom cabinets are available only in width increments of 3". Custom cabinets are the most expensive type. They can be built in any width to fit a room exactly. Almost any style, size, or shape is possible.
2. This cabinet would be a wall cabinet that is 27" wide and 30" tall.
3. A single, center-mounted guide is located along the centerline of the drawer. It can be used only on cabinetry that has a face frame. Side-mounted guides are stronger because they support both sides of the drawer. They can be used on face-frame or frameless cabinets. They allow the drawer to extend either partially or fully beyond the front of the cabinet.
4. Cup hinges eliminate the need for a face frame. They can be adjusted in several planes (up/down, side-to-side, in/out) very easily. They are quite strong, and they can be installed quickly. A cup hinge is not visible for the outside of the cabinet.

■ Section 36.2 On the Job
Answers will vary.

■ Section 36.3 Check Your Knowledge

1. While the limits for counter height range from 30" to 38", the standard height in a kitchen is 36".
2. During the manufacture of plastic laminate, decorative surface papers are bonded to other materials. Because the surface papers can be printed before assembly, plastic laminate comes in a large number of colors and patterns.
3. This type of countertop is easy to install because it already incorporates both the backsplash and the countertop edge. Also, the laminate is already in place.
4. A piece of cardboard or hardboard is scribed to fit the wall, then the pattern is transferred to the countertop so it can be trimmed.

■ Section 36.3 On the Job
Answers will vary.

■ Section 36.3 Estimating on the Job
60 square feet
30 lineal feet

■ Chapter 36 Review Questions

1. The installation of cabinetry occurs just before interior trim is installed, or sometimes at the same time. This is usually after the finish floor is in place.
2. The five layouts are U shape, L shape, parallel wall, side wall, and island.
3. The work triangle represents the shortest walking distance in a kitchen between the refrigerator, the primary cooking surface, and the sink.
4. Bathroom base cabinets are usually 30" high and 21" deep.
5. A face frame fits around the front opening in the carcase and provides a mounting surface for hinges and drawer hardware. Frameless cabinets have no face frames.
6. The two basic types are inset, or flush, and overlay doors. Inset doors fit entirely within the door opening. A small gap is required between the door and the face frame to provide clearance. These are sometimes referred to as flush doors. Overlay doors fit over the edge of the carcase or face frame.
7. Locate the highest part of the floor in the area of the cabinets. This is the starting point for the layout.
8. Use at least #10 round head screws that are long enough to go through the ¾" back rail and the wall covering and extend at least 1" into the studs.
9. If enough material is available on the back of the cabinet, the cabinet should be held in place, scribed, and cut to fit the irregular wall surface.
10. Contact cement should be used. The adhesive should be applied to both mating surfaces (the back of the laminate and the surface of the substrate) in an even coating, using a brush or a roller.

Instructional Plan

Focus & Planning

Objectives
- Identify the three basic types of paneling.
- Explain how paneling should be stored and conditioned.
- Explain how to install sheet paneling.
- Estimate the amount of sheet paneling required for a room.

SAFETY FIRST
- ❑ Review the safety rules in the chapter.
- ❑ Make sure all necessary personal protective equipment is available and in good condition.
- ❑ Make sure all required tools and equipment are available and in good working order.

Instruction & Student Practice

Using the *Student Textbook*, have students:
- ❑ Read Section 37.1.
- ❑ Complete Section 37.1 Check Your Knowledge and On the Job.
- ❑ Read Section 37.2.
- ❑ Complete Section 37.2 Check Your Knowledge and On the Job.
- ❑ Read Section 37.3.
- ❑ Read Estimating Paneling and complete Section 37.3 Estimating on the Job.
- ❑ Complete Section 37.3 Check Your Knowledge and On the Job.

Using *Carpentry Applications*, assign:
- ❑ 37-1: Estimating Wood Paneling
- ❑ 37-2: Installing the First Wall Panel

Using *Carpentry Math*, assign:
- ❑ M37-1: Calculating Remodeling Bids

Using the *Safety Guidebook*, assign:
- ❑ 2-1: Musculoskeletal Disorders
- ❑ 5-14: Table Saws

Review & Student Performance

- ❑ Assign Review Questions from the end of the chapter.
- ❑ Assign the Chapter 37 Test or create a test using the *ExamView® Pro Test Generator* on the *Instructor Productivity CD-ROM*.
- ❑ Create and administer a Unit 7 Posttest using the *ExamView® Pro Test Generator* on the *Instructor Productivity CD-ROM*.
- ❑ Use *Carpentry Applications* job sheets as desired for final performance assessment.

(Continued on next page)

Carpentry & Building Construction Instructor Resource Guide
Copyright © Glencoe/McGraw-Hill

Answers to Textbook Questions

Section 37.1 Check Your Knowledge

1. Wainscoting is paneling that runs partway up the wall from the floor.
2. Lap joints or tongue-and-groove joints are used with board paneling.
3. Paneling should be stacked on the floor with stickers between layers.
4. This allows panels to acclimate to the temperature and humidity of the room.

Section 37.1 On the Job

Answers will vary.

Section 37.2 Check Your Knowledge

1. Drywall is recommended. It provides a fire-resistant base and solid support.
2. It is not necessary when panels are attached with adhesives.
3. Position the panel at the proper height by shimming it. When the panel is set perfectly plumb and at the correct height, set a compass for the widest gap between the panel and the corner. Scribe a line down the edge of the panel.
4. The front edge of any outlet box should be flush with the surface of the wood paneling when the job is complete. If necessary, the box should be repositioned or fitted with a box extender.

Section 37.2 On the Job

84 lineal feet

Section 37.3 Estimating on the Job

553 board feet

Section 37.3 Check Your Knowledge

1. A nominal 8" width is the maximum recommended for most parts of the country.
2. For paneling that is to be installed directly to studs, adequate blocking should be placed between the studs to provide nailing support. The blocking should not be more than 24" OC.
3. In blind nailing the nails are driven at an angle through the tongue of the board. This allows a subsequent board to conceal the nails.
4. Blocking is not required because the boards can span the distance between studs.

Section 37.3 On the Job

(1) Height and length of areas to be paneled: 12' × 3' and 14.5' × 3'. (2) The face width of a 1 × 4 T&G board is 3 $\frac{3}{16}$" or 3.1875". (3) Convert the areas to inches: 12' × 12 = 144"; 14.5' × 12 = 174"; 144" + 174" = 318". (4) Divide by the face width: 318" ($3.1875 = 99.76$). Round to 100 boards. (5) Multiply number of boards by their length: 100 × 4 = 400. (6) You'll need 400 lineal feet of paneling, and you'll have 12" waste on each board.

Chapter 37 Review Questions

1. Paneling that runs only part of the way up the wall is called wainscoting. It is usually about 32" high.
2. The three basic types of paneling are
   - Sheet paneling is made of plywood, hardboard, or medium-density fiberboard (MDF). It is most commonly available in 4 × 8 sheets. Thicknesses range from $\frac{5}{32}$" to $\frac{3}{4}$". Edges along the length of the panels may be square or rabetted. Edges along the width are square.
   - Board paneling is made of solid wood, including Douglas fir, oak, and various species of pine. It ranges from $\frac{3}{8}$" to $\frac{3}{4}$" thick and comes in lengths of 8' to 12' or more. The edges of each board interlock with adjoining boards in either lap joint or tongue-and-groove joint.
   - Raised paneling is constructed much like raised-panel cabinet doors and is made of solid wood, particularly oak and cherry. Individual raised panels are held in place by a grid of stiles and rails secured to the walls with nails or screws.
3. Paneling should always be stored indoors. It should be stacked on the floor, ideally with stickers between the sheets to prevent warping. If panels must be stored on edge, rest them on a long edge.
4. MDF panels are susceptible to damage caused by moisture and high humidity. They should not be installed in unheated rooms or in humid areas such as basements and bathrooms. Panels $\frac{5}{32}$" thick should always be installed over a noncombustible backing such as drywall, and should never be installed over masonry.
5. There should be a ¼" clearance at top and bottom of each panel to allow for expansion and contraction.
6. The panels should not be butted together tightly. A $\frac{1}{16}$" gap is recommended for hardboard and MDF panels to allow for expansion. A smaller gap is recommended for plywood panels.
7. Be sure to follow the adhesive manufacturer's instructions. After the panels are properly cut and fitted, the adhesive is applied to the wall surface with a caulking gun in a continuous $\frac{1}{8}$" wide bead around the perimeter of the panel and around any cutouts. Apply additional adhesive in a zigzag pattern in the middle.
8. Panels that are approved for use over masonry are typically installed over furring strips. The wall must be waterproofed first. Where extreme humidity may cause condensation on the inside of an exterior masonry wall, a vapor barrier should be used to prevent moisture penetration to the panel.
9. It is called a herringbone, or chevron, pattern.
10. The allowance for waste is 5 percent.

Instructional Plan

Chapter 38: Mechanicals

Focus & Planning

Objectives
- Describe or sketch a simple plumbing system.
- Recognize the various types of piping used for water supply and the DWV system.
- Describe the basic elements of an electrical system.
- Understand how split-system air conditioners work and identify the basic parts of the system.

SAFETY FIRST
- ❏ Review the safety rules in the chapter.
- ❏ Make sure all necessary personal protective equipment is available and in good condition.
- ❏ Make sure all required tools and equipment are available and in good working order.

Instruction & Student Practice

Using the *Student Textbook*, have students:
- ❏ Read Section 38.1.
- ❏ Complete Section 38.1 Check Your Knowledge and On the Job.
- ❏ Read Section 38.2.
- ❏ Complete Section 38.2 Check Your Knowledge and On the Job.
- ❏ Read Section 38.3.
- ❏ Complete Section 38.3 Check Your Knowledge and On the Job.

Using *Carpentry Applications*, assign:
- ❏ 38-1: Roughing-In Electrical Wiring
- ❏ 38-2: Installing Electrical Switches and Receptacles

Using *Carpentry Math*, assign:
- ❏ M38-1: Estimating Plumbing Materials
- ❏ M38-2: Measuring Electricity
- ❏ M38-3: Estimating Heat Loss

Using the *Safety Guidebook*, assign:
- ❏ 2-2: Personal Protective Equipment
- ❏ 3-2: Electrical Safety
- ❏ 5-1 to 5-26. These relate to safe use of tools and equipment.

Using the *Instructor Productivity CD-ROM*, show PowerPoint presentation:
- ❏ 38-1: Roughing-in Electrical Wiring
- ❏ 38-2: Installing Electrical Switches and Receptacles

Review & Student Performance

- ❏ Assign Review Questions from the end of the chapter.
- ❏ Assign the Chapter 38 Test or create a test using the *ExamView® Pro Test Generator* on the *Instructor Productivity CD-ROM*.
- ❏ Create and administer a Unit 8 Pretest using the *ExamView® Pro Test Generator* on the *Instructor Productivity CD-ROM*.
- ❏ Use *Carpentry Applications* job sheets as desired for final performance assessment.

(Continued on next page)

Answers to Textbook Questions

■ Section 38.1 Check Your Knowledge

1. The abbreviation stands for heating, ventilation, and air conditioning.
2. The plumbing inspector must check the plumbing installation at the rough-in stage and the finish stage.
3. A trap is a curved section of drainpipe that is located beneath a fixture. It prevents sewer gases in the waste pipes from entering the house but does not block drainage. A small amount of water in the bottom of each trap serves as a plug.
4. They can be reinforced by nailing a 2× scab to each side.

■ Section 38.1 On the Job

Answers will vary.

■ Section 38.2 Check Your Knowledge

1. An ampere (amp) is a measure of electrical current.
2. The circuit breaker is like a fast-acting switch. It shuts off power in a circuit if it detects overload conditions that might lead to a fire.
3. A receptacle has a combination of slots and grounding holes sized to accept the prongs of an electrical plug.
4. Inside each outlet box, the outer sheathing of the cable is stripped off to expose individual conductors. The conductors are left exposed until the final wiring stage.

■ Section 38.2 On the Job

Answers will vary.

■ Section 38.3 Check Your Knowledge

1. Four types of heating fuel are oil, electricity, natural gas, and solar.
2. As air within the room cools, it sinks to the floor and flows into the return air register. Return ducts carry the cooled air back to the furnace. There, it is reheated and recirculated.
3. The fins maximize the transfer of heat to the surrounding air.
4. The air handler draws in warmed house air through ducts and blows it over the evaporator coils. Refrigerant in the coils absorbs heat from the air. The cooled air is then distributed to the house.

■ Section 38.3 On the Job

Answers will vary.

■ Chapter 38 Review Questions

1. The three types of pipes are supply pipes, waste pipes, and vent pipes.
2. Supply pipes are pressurized. The DWV side of a plumbing system is not pressurized.
3. The middle third of a floor joist must not be notched.
4. One end of a cast iron pipe (the bell end) is flared. The other end (the spigot end) fits into the flared end. The joints are then sealed.
5. A circuit is a cable or group of cables that supplies electricity to a specific area of the house.
6. *Any three:* They provide a convenient location for joining wire. They contain short circuits and other wiring faults that could cause fires. They provide a solid mounting surface for switches and other devices. They prevent dust and debris from accumulating around wiring connections.
7. The water is heated in the boiler.
8. Rather than heating air, as in a forced-air or hydronic system, a radiant system heats a material, which then radiates the heat into the room.
9. The parts include refrigerant coils, air handler, and compressor.
10. An HRV (heat recovery ventilator) extracts the heat from stale indoor air before exhausting the air outdoors. The heat is then transferred to fresh air drawn into the house.

Instructional Plan

Focus & Planning

Objectives
- Identify several types of insulation.
- Interpret an insulator's R-value in determining its effectiveness as insulation.
- Identify the best uses for common types of insulating materials.
- Explain the importance of vapor barriers and ventilation.
- Describe several types of wall construction that reduce noise transmission.

SAFETY FIRST
- ❏ Review the safety rules in the chapter.
- ❏ Make sure all necessary personal protective equipment is available and in good condition.
- ❏ Make sure all required tools and equipment are available and in good working order.

Instruction & Student Practice

Using the *Student Textbook*, have students:
- ❏ Read Section 39.1.
- ❏ Complete Section 39.1 Check Your Knowledge and On the Job.
- ❏ Read Section 39.2.
- ❏ Complete Section 39.2 Check Your Knowledge and On the Job.

Using *Carpentry Applications*, assign:
- ❏ 39-1: Estimating Fiberglass Insulation
- ❏ 39-2: Installing Fiberglass Insulation

Using *Carpentry Math*, assign:
- ❏ M39-1: Estimating Insulation

Using the *Safety Guidebook*, assign:
- ❏ 2-2: Personal Protective Equipment
- ❏ 6-3: Hazardous Materials

Using the *Instructor Productivity CD-ROM*, show PowerPoint presentation:
- ❏ 39-2: Installing Fiberglass Insulation

Review & Student Performance

- ❏ Assign Review Questions from the end of the chapter.
- ❏ Assign the Chapter 39 Test or create a test using the *ExamView® Pro Test Generator* on the *Instructor Productivity CD-ROM*.
- ❏ Use *Carpentry Applications* job sheets as desired for final performance assessment.

(Continued on next page)

Answers to Textbook Questions

■ Section 39.1 Check Your Knowledge
1. The insulated areas must be covered by a plastic film vapor barrier.
2. 2×4 walls contain R-11 insulation. 2×6 walls contain R-19 insulation.
3. This is a continuous insulation layer that separates conditioned space from areas exposed to outdoor temperatures.
4. Emissivity is a material's ability to radiate heat.

■ Section 39.1 On the Job
1,120 − 168 = 952 sq. ft.
952 ÷ 60 = 15.86, or 16 batts.

■ Section 39.2 Check Your Knowledge
1. The wall rated STC 55 would be best at reducing sound transmission.
2. It is a gap that allows sound to get around a material without actually going through it.
3. Even a hole of 1 square inch in a wall rated at STC 50 can reduce that wall's performance to STC 30.
4. The INR rating is based on decibels.

■ Section 39.2 On the Job
Answers will vary.

■ Chapter 39 Review Questions
1. Climate is the primary factor to consider when choosing thermal insulation.
2. The four types are flexible batt, loose-fill, rigid sheet, and spray-foam.
3. R-value is a measure of a material's ability to resist heat transmission.
4. Loose-fill insulation is often used in attic floors where HVAC pipes and wiring make it difficult to install batt insulation.
5. Extruded polystyrene (XEPS) would be suitable for insulating the outside of a foundation wall below grade.
6. A vapor barrier is a material highly resistant to vapor transmission.
7. Attic ventilation removes moisture vapor, reduces the chance of ice dams forming in the winter, and lowers the temperature of the attic in the summer.
8. Radiant heat travels in a straight line away from a hot surface and heats anything solid it meets.
9. See Figs. 39-21 (C, D, or E) and 39-23.
10. Sound-insulating materials block noise transmission. Sound-absorbing materials stop the reflection of sound back into a room.

Instructional Plan

Chapter 40: Walls and Ceilings

Focus & Planning

Objectives
- Identify and describe the various types of drywall.
- Describe a nail pop and explain the methods used to prevent it.
- Describe problems relating to safety and health when installing drywall and explain preventive measures.
- Identify the basic materials used in three-coat plaster work.
- Install a suspended ceiling.

SAFETY FIRST
- ❏ Review the safety rules in the chapter.
- ❏ Make sure all necessary personal protective equipment is available and in good condition.
- ❏ Make sure all required tools and equipment are available and in good working order.

Instruction & Student Practice

Using the *Student Textbook*, have students:
- ❏ Read Section 40.1.
- ❏ Complete Section 40.1 Check Your Knowledge and On the Job.
- ❏ Read Section 40.2.
- ❏ Read Estimating Gypsum Lath, Nails, and Labor and Complete Section 40.2 Estimating On the Job.
- ❏ Complete Section 40.2 Check Your Knowledge and On the Job.
- ❏ Read Section 40.3.
- ❏ Complete Section 40.3 Check Your Knowledge and On the Job.

Using *Carpentry Applications*, assign:
- ❏ 40-1: Hanging Drywall
- ❏ 40-2: Taping and Finishing Drywall

Using *Carpentry Math*, assign:
- ❏ M40-1: Estimating Drywall Insulation Materials

Using the *Safety Guidebook*, assign:
- ❏ 2-2, 6-4. These relate to personal safety.
- ❏ 5-3: Edge-Cutting and Shaping Tools
- ❏ 6-3: Hazardous Materials

Using the *Instructor Productivity CD-ROM*, show PowerPoint presentation:
- ❏ 40-1: Hanging Drywall
- ❏ 40-2: Taping and Finishing Drywall

Review & Student Performance

- ❏ Assign Review Questions from the end of the chapter.
- ❏ Assign the Chapter 40 Test or create a test using the *ExamView® Pro Test Generator* on the *Instructor Productivity CD-ROM*.
- ❏ Use *Carpentry Applications* job sheets as desired for final performance assessment.

(Continued on next page)

Answers to Textbook Questions

■ Section 40.1 Check Your Knowledge
1. In general, a drywall panel is stronger in the long dimension.
2. Both *Type X* and *Type C* are fire-code drywall.
3. When fast curing is important, setting-type compound would be used.
4. Use respiratory protection. Use a pole sander and a vacuum-based sanding system instead of sanding by hand.

■ Section 40.1 On the Job
The ceiling has an area of 156 sq. ft. and will require 6 panels.

■ Section 40.2 Estimating on the Job
It needs 6 bundles of lath and 20 lbs. of nails. Labor would cost $1050.

■ Section 40.2 Check You Knowledge
1. Gypsum lath, metal lath, and wood lath are used as a base for plaster.
2. A plaster ground is a material permanently or temporarily attached to a surface to be plastered. It provides a straight edge and helps the plasterer gauge the thickness of the installation.
3. Cornerite is a length of metal lath that is used to strengthen interior corners.
4. The two finishes are the sand-float and the putty finish.

■ Section 40.2 On the Job
Answers will vary.

■ Section 40.3 Check Your Knowledge
1. The advantages of suspended ceilings are the following:
   - A suspended ceiling covers bare joists, exposed pipes, and wiring and may be used to lower a high ceiling.
   - Access to valves, switches, and controls hidden by the suspended ceiling is possible because the panels can be removed easily.
   - The ceilings also reduce noise between floors.
2. The grid system includes main beams, cross tees, wall molding, and ceiling panels.
3. A suspended ceiling grid is a network of hanger wires attached to the ceiling joists with hanger-wire screws.
4. The border panels must be cut because they must fit between the first row of cross tees and the walls.

■ Section 40.3 On the Job
Answers will vary.

■ Chapter 40 Review Questions
1. Tapering allows the joints between panels to be filled and taped.
2. It is required by code on the outer surface of walls separating an attached garage from the house.
3. Moisture-resistant drywall, also called MR drywall or green board, should be used.
4. Nail pops are greatly reduced if the moisture content of the framing is less than 15 percent when the drywall is applied. The use of screws nearly eliminates the problem.
5. Feathering is a process of smoothing the edges of joint compound so that there are no ridges.
6. Joint compound contains silica.
7. Water-resistant sheathing paper is placed behind the metal lath.
8. Plaster is made from combinations of sand, lime or prepared plaster, and water.
9. The first plaster coat is the scratch coat. The second coat is the brown, or leveling, coat. The third coat is the finish coat.
10. Placement of the first main beam and cross tees is marked with strings.

Instructional Plan

Focus & Planning

Objectives
- Describe the differences between the two basic types of finish.
- List the basic ingredients of paint.
- Give the steps in painting a house exterior.
- Give the steps in painting an interior.
- Diagnose problems with painted finishes.

SAFETY FIRST
- ❏ Review the safety rules in the chapter.
- ❏ Make sure all necessary personal protective equipment is available and in good condition.
- ❏ Make sure all required tools and equipment are available and in good working order.

Instruction & Student Practice

Using the *Student Textbook*, have students:
- ❏ Read Section 41.1.
- ❏ Complete Section 41.1 Check Your Knowledge and On the Job.
- ❏ Read Section 41.2.
- ❏ Complete Section 41.2 Check Your Knowledge and On the Job.
- ❏ Read Section 41.3.
- ❏ Read Estimating Interior Paint Needs and complete Section 41.3 Estimating on the Job.
- ❏ Complete Section 41.3 Check Your Knowledge and On the Job.

Using *Carpentry Applications*, assign:
- ❏ 41-1: Painting Ceilings and Walls
- ❏ 41-2: Cleaning a Paintbrush

Using *Carpentry Math*, assign:
- ❏ M41-1: Estimating House Paint
- ❏ M41-2: Estimating Flooring Materials

Using the *Safety Guidebook*, assign:
- ❏ 4-1: Ladders
- ❏ 4-2: Scaffolds
- ❏ 6-3: Hazardous Materials

Review & Student Performance

- ❏ Assign Review Questions from the end of the chapter.
- ❏ Assign the Chapter 41 Test or create a test using the *ExamView® Pro Test Generator* on the *Instructor Productivity CD-ROM*.
- ❏ Use *Carpentry Applications* job sheets as desired for final performance assessment.

(Continued on next page)

Answers to Textbook Questions

Section 41.1 Check Your Knowledge

1. Pigments protect wood from ultraviolet (UV) rays. They also add color.
2. A binder is a resin that binds particles of pigment together. It forms a film after the carrier evaporates.
3. The binders in oil-base paints are suspended in a mineral spirit carrier. The binders in latex paints are suspended in water.
4. A primer is not good at blocking UV radiation, so it must be covered with two coats of standard paint.

Section 41.1 On the Job

Answers will vary.

Section 41.2 Check Your Knowledge

1. Finish problems can be expensive to correct. It is therefore wise to use only top-quality products.
2. Boxing paint means to pour paint back and forth from one can into another. This evens out any slight variations in color.
3. A narrow sash brush is used to paint the mullions of the window.
4. To protect bristles as they dry, wrap brushes in heavy paper or a cardboard sheath and lay them in a dry location. Some painters hang brushes to dry. Roller covers should be stored on end so that the nap is not flattened.

Section 41.2 On the Job

Answers will vary.

Section 41.3 Estimating on the Job

The room has 378 sq. ft. One coat of primer will require 3 qts. Two coats of finish paint will require 1 ½ gals. Door and windows will require 1 qt. of primer and 1 qt. finish paint.

Section 41.3 Check Your Knowledge

1. Sheen describes the shininess of a surface.
2. Edging is using a brush to paint into the corners between large flat surfaces, where a roller cannot reach.
3. Hold a cardboard, metal, or plastic guard flush against the bottom edge of the baseboard or mask off the floor to prevent the brush from picking up dirt.
4. When painting a raised panel door, paint the panel molding first and then the panels themselves.

Section 41.3 On the Job

Answers will vary.

Chapter 41 Review Questions

1. Film-forming finishes coat the wood surface. Penetrating finishes soak into the wood.
2. All paints contain pigments, binders, and carriers.
3. A primer is a paint that has a higher proportion of binder than standard paint. This enables it to hold particularly well to unpainted wood surfaces.
4. Penetrating finishes are very easy to apply. They do not obscure the wood grain.
5. Engineered wood behaves predictably. Its performance is uniform. Products are dimensionally stable over a wide range of widths and thicknesses. There are no defects such as those commonly found in solid lumber.
6. The top of a two-story house is painted first.
7. The outdoor temperature must stay above 40°F [4°C] for at least twenty-four hours after applying oil-base paints. It must stay above 50°F [10°C] for at least twenty-four hours after applying latex paints.
8. The previous paint film may have been applied in several heavy coats without sufficient drying time between coats. The primer may not be compatible with the finish coat.
9. The five types of interior paints based on sheen are enamel, semi-gloss enamel, pearl, eggshell, flat.
10. Always start with the ceiling, and then paint the walls. Complete the job by painting wood trim and doors.

Instructional Plan

Focus & Planning

Objectives
- List the three most common forms of wood flooring.
- Identify tree species used to create hardwood flooring.
- Secure wood flooring to a plywood subfloor.
- Describe two methods for applying wood flooring to a concrete subfloor.

SAFETY FIRST
- ❏ Review the safety rules in the chapter.
- ❏ Make sure all necessary personal protective equipment is available and in good condition.
- ❏ Make sure all required tools and equipment are available and in good working order.

Instruction & Student Practice

Using the *Student Textbook*, have students:
- ❏ Read Section 42.1.
- ❏ Complete Section 42.1 Check Your Knowledge and On the Job.
- ❏ Read Section 42.2.
- ❏ Read Estimating Strip Flooring and complete Section 42.2 Estimating on the Job.
- ❏ Complete Section 42.2 Check Your Knowledge and On the Job.

Using *Carpentry Applications*, assign:
- ❏ 42-1: Laying Engineered-Wood Flooring
- ❏ 42-2: Estimating Strip Flooring

Using *Carpentry Math*, assign:
- ❏ M42-1: Estimating Flooring Materials

Using the *Safety Guidebook*, assign:
- ❏ 2-2: Personal Protective Equipment
- ❏ 5-1: Hand Tools
- ❏ 6-3: Hazardous Materials

Review & Student Performance

- ❏ Assign Review Questions from the end of the chapter.
- ❏ Assign the Chapter 42 Test or create a test using the *ExamView® Pro Test Generator* on the *Instructor Productivity CD-ROM*.
- ❏ Use *Carpentry Applications* job sheets as desired for final performance assessment.

(Continued on next page)

Answers to Textbook Questions

Section 42.1 Check Your Knowledge

1. Grades are clear, select, No. 1 Common, No. 2 Common, and shorts.
2. Beech, birch, and maple are graded using the same system. Oak is graded using a different system.
3. If flooring absorbs moisture from the subfloor, the underside of the flooring will expand. This causes boards to cup slightly.
4. Engineered wood flooring is made of three, five, seven, or more layers of wood veneer or thin wood strips that have been bonded together.

Section 42.1 On the Job

Answers will vary.

Section 42.2 Estimating on the Job

The job requires 209.5 bd. ft. of flooring. It will take 5 hours to lay, 2 hours to sand, and 3.8 hours to finish.

Section 42.2 Check Your Knowledge

1. A power nailer must be struck by a mallet.
2. The paper protects flooring from moisture that might come from below and helps to prevent squeaks.
3. A prybar should be used to lever floorboards into place. Chisels and screwdrivers should never be used.
4. Many cities restrict the use of floor finishes containing VOCs.

Section 42.2 On the Job

Answers will vary.

Chapter 42 Review Questions

1. A plank is any solid-wood flooring board that is at least 3" wide. A flooring strip is a solid-wood flooring board that is no more than 3 ½" wide.
2. This is a process of driving fasteners at an angle through the edge of each board, so that the fasteners will be concealed by the next board.
3. Parquet flooring refers to any flooring assembled with small, precisely cut pieces of wood into a geometric pattern. Prefabricated parquet tiles consist of individual pieces of solid wood that are factory-adhered to a plywood backing.
4. The three methods used to install engineered wood flooring are the nail-down, floating, and glue-down methods.
5. Oak, maple, beech, birch, cherry, walnut, and plantation-grown tropical hardwoods are used to create hardwood flooring.
6. The process is acclimation.
7. *Any three of the following:*
   - Never unload wood flooring when it is raining or snowing.
   - Cover flooring with a tarp in foggy or damp conditions.
   - Store flooring in a well-ventilated and weathertight building.
   - Do not store or lay flooring in a damp building.
   - Do not lay flooring in a cold building.
8. The first course can be measured to a string stretched between two nails along the length of the room. Some installers snap a chalk line on the floor instead of measuring to a string.
9. Concrete gives off moisture as it cures, which can cause wood flooring to cup.
10. A sleeper is a length of lumber that supports wood flooring over concrete.

Instructional Plan

Chapter 4: Resilient Flooring & Ceramic Tile

Focus & Planning

Objectives
- Describe the basic method for laying sheet vinyl flooring.
- Estimate the quantity of resilient flooring needed for any room.
- Identify the basic tools used for installing ceramic tile.
- Identify a ceramic tile by its characteristics.

SAFETY FIRST
- ❏ Review the safety rules in the chapter.
- ❏ Make sure all necessary personal protective equipment is available and in good condition.
- ❏ Make sure all required tools and equipment are available and in good working order.

Instruction & Student Practice

Using the *Student Textbook*, have students:
- ❏ Read Section 43.1.
- ❏ Complete Section 43.1 Check Your Knowledge and On the Job.
- ❏ Read Section 43.2.
- ❏ Complete Section 43.2 Check Your Knowledge and On the Job.

Using *Carpentry Applications*, assign:
- ❏ 43-1: Installing Vinyl Flooring
- ❏ 43-2: Cutting Ceramic Tile

Using *Carpentry Math*, assign:
- ❏ M43-1: Estimating Ceramic Tile

Using the *Safety Guidebook*, assign:
- ❏ 2-2: Personal Protective Equipment
- ❏ 5-1: Hand Tools
- ❏ 6-3: Hazardous Materials

Review & Student Performance

- ❏ Assign Review Questions from the end of the chapter.
- ❏ Assign the Chapter 43 Test or create a test using the *ExamView® Pro Test Generator* on the *Instructor Productivity CD-ROM*.
- ❏ Use *Carpentry Applications* job sheets as desired for final performance assessment.

(Continued on next page)

Answers to Textbook Questions

Section 43.1 Check Your Knowledge

1. All resilient flooring is flexible, springy, and less than 3/16" thick.
2. Vinyl is synthetic.
3. Mastic is used to install resilient flooring.
4. These chalk lines identify the center and the width of the seam area.

Section 43.1 On the Job

627 tiles are needed without waste. Including five percent waste, the total is 659 tiles.

Section 43.2 Check Your Knowledge

1. A bisque is a piece of tile minus the glaze.
2. The four types of tile are nonvitreous, semi-vitreous, vitreous, and impervious tile.
3. Mosaic tile is any tile that is no larger than 2" square.
4. Sanded grout is simply plain grout to which sand has been added. This improves its strength. Sanded grout is used for joints wider than 1/16".

Section 43.2 On the Job

Answers will vary.

Chapter 43 Review Questions

1. Underlayment prevents small flaws in the subfloor from showing through. It also provides firm, clean, void-free support.
2. The quantity of sheet vinyl flooring is estimated by the square footage of the room, plus an allowance for waste, seams, and pattern matching.
3. Houses in hot climates are often built on a concrete slab foundation. This type of foundation serves as an excellent base for the installation of ceramic tile. Also, tile is not affected by humid conditions.
4. Wall tile is generally a nonvitreous tile with a relatively soft glaze. Floor tile can be any kind of tile (from nonvitreous to impervious, glazed or unglazed) that is strong enough to hold up in use on the floor.
5. A lugged tile has spacing lugs built into its sides. When the tiles are placed edge-to-edge, the lugs automatically determine the proper spacing between them.
6. Tile substrate made with cement is called backerboard or cement board. It is 1/2" or 5/8" thick.
7. A nibbler is used to cut shapes in ceramic tile by "nibbling" at the edges.
8. Grout is a form of mortar used to fill spaces between tiles. It prevents moisture and dirt from getting between the tiles.
9. Keep other electrical tools away from the area in which the wet saw is being used. Make sure all power tools, including the wet saw, are plugged into GFCI-protected circuits.
10. The three layers of mortar are the scratch coat, the bed, and the bond coat.

Instructional Plan

CHAPTER 44: Chimneys and Fireplaces

Focus & Planning

Objectives
- Identify the main parts of a chimney.
- Identify the main parts of a fireplace.
- Understand how chimneys and fireplaces are installed to limit fire hazards.

SAFETY FIRST
- ❑ Review the safety rules in the chapter.
- ❑ Make sure all necessary personal protective equipment is available and in good condition.
- ❑ Make sure all required tools and equipment are available and in good working order.

Instruction & Student Practice

Using the *Student Textbook*, have students:
- ❑ Read Section 44.1.
- ❑ Complete Section 44.1 Check Your Knowledge and On the Job.
- ❑ Read Section 44.2.
- ❑ Complete Section 44.2 Check Your Knowledge and On the Job.

Using *Carpentry Applications*, assign:
- ❑ 44-1: Planning for a Prefabricated Fireplace
- ❑ 44-2: Installing the Front Hearth and Mantel

Using *Carpentry Math*, assign:
- ❑ M44-1: Estimating Brick Courses

Using the *Safety Guidebook*, assign:
- ❑ 2-1: Musculoskeletal Disorders
- ❑ 4-2: Scaffolds

Review & Student Performance

- ❑ Assign Review Questions from the end of the chapter.
- ❑ Assign the Chapter 44 Test or create a test using the *ExamView® Pro Test Generator* on the *Instructor Productivity CD-ROM*.
- ❑ Use *Carpentry Applications* job sheets as desired for final performance assessment.

(Continued on next page)

Answers to Textbook Questions

■ Section 44.1 Check Your Knowledge

1. Draft creates air movement that draws air into the fuel-burning appliance to aid in combustion and expels smoke and harmful gases.
2. *Any two*: The dimensions, height, shape, and smoothness of the flue determine the effectiveness of the chimney in producing adequate draft.
3. There must be 2" of clearance between the chimney walls and wood framing members.
4. Cement mortar is more resistant to the action of heat and flue gases than lime mortar.

■ Section 44.1 On the Job

Answers will vary.

■ Section 44.2 Check Your Knowledge

1. Makeup air is air drawn into a house to replace air exhausted by a combustion appliance.
2. A front hearth should be 16" wide, extend at least 8" on either side of the fireplace opening, and be at least 2" thick.
3. The front of the smoke chamber is corbeled.
4. Woodwork can come no closer than 6" from the firebox opening.

■ Section 44.2 On the Job

Answers will vary.

■ Chapter 44 Review Questions

1. A chimney in the interior of a house will have better draft because there will be greater difference in temperature between chimney gases and outside atmosphere.
2. The main parts of a chimney include the foundation, flue liners, walls, and cleanout opening.
3. Refractory cement should be used to join flue sections.
4. Building codes generally require a separate flue for each fireplace, furnace, and boiler.
5. A corbel is a course of brick offset to extend past the course below it.
6. A chimney cap prevents moisture from seeping between the brick and flue liner. It also drains water away from the top of the chimney.
7. The main parts of a fireplace include the firebox, hearth, lintel and throat, damper, smoke shelf, and smoke chamber.
8. The relationships among the depth, height, and width of the firebox are the most important factors for the proper operation of a fireplace.
9. In cold weather, closing the damper reduces heat loss when the chimney is not in use. In warm weather, a closed damper prevents insects and small animals from entering the house.
10. A zero-clearance fireplace is a prefabricated fireplace that can be placed directly against wood framing.

Instructional Plan

Focus & Planning

Objectives
- Name the basic types of materials used for decking.
- Name the basic various elements of a deck.
- Lay out piers for a rectangular attached deck.
- Plumb a post.
- Handle and cut preservative-treated wood safely.
- Describe two methods for installing concrete porch steps.

SAFETY FIRST
- ❑ Review the safety rules in the chapter.
- ❑ Make sure all necessary personal protective equipment is available and in good condition.
- ❑ Make sure all required tools and equipment are available and in good working order.

Instruction & Student Practice

Using the *Student Textbook*, have students:
- ❑ Read Section 45.1.
- ❑ Complete Section 45.1 Check Your Knowledge and On the Job.
- ❑ Read Section 45.2.
- ❑ Complete Section 45.2 Check Your Knowledge and On the Job.

Using *Carpentry Applications*, assign:
- ❑ 45-1: Building a Deck, Part 1
- ❑ 45-2: Building a Deck, Part 2

Using *Carpentry Math*, assign:
- ❑ M45-1: Estimating Support Posts for Decks

Using the *Safety Guidebook*, assign:
- ❑ 2-2: Personal Protective Equipment
- ❑ 6-3: Hazardous Materials

Review & Student Performance

- ❑ Assign Review Questions from the end of the chapter.
- ❑ Assign the Chapter 45 Test or create a test using the *ExamView® Pro Test Generator* on the *Instructor Productivity CD-ROM*.
- ❑ Create and administer a Unit 8 Posttest using the *ExamView® Pro Test Generator* on the *Instructor Productivity CD-ROM*.
- ❑ Use *Carpentry Applications* job sheets as desired for final performance assessment.

(Continued on next page)

Answers to Textbook Questions

■ Section 45.1 Check Your Knowledge

1. The structure of the house and a system of concrete piers support an attached deck.
2. The three basic types are redwood, cedar, and preservative-treated southern yellow pine.
3. A preservative called CCA (chromated copper arsenate) is being phased out of use.
4. Galvanizing makes the metal more corrosion resistant.

■ Section 45.1 On the Job

Answers will vary.

■ Section 45.2 Check Your Knowledge

1. Built-up beams can be assembled into a beam of any length. Another advantage is that a built-up beam is easier to position because it can be assembled in place.
2. The cut ends of solid or built-up beams should be coated with a water repellent.
3. Install metal flashing to prevent water from rotting nearby wood. Use lag bolts or through bolts to connect the ledger to studs, plates, or rim joists.
4. Blind nailing prevents the fasteners from showing. It also reduces the chance that water will get into the wood around fastener heads.

■ Section 45.2 On the Job

Exactly 0.6984 cubic yard is required. Round up to ¾ cubic yard.

■ Chapter 45 Review Questions

1. Decks must be weather resistant, safe, and strong.
2. Softwood decking is typically 1 ¼" or 1 ⅝" thick.
3. The four characteristics are appearance, strength, moisture content, and decay resistance.
4. Heartwood is the portion of a tree nearest the core. It is typically dark in color and is the most decay-resistant wood. Sapwood is the outer growth layer of the tree. It is lighter in color than heartwood. It is less decay-resistant than heartwood.
5. A pier is a concrete column that supports a concentrated load such as a post.
6. Cut preservative-treated wood over a tarp. Dispose of the collected sawdust as directed by local regulations.
7. Softwoods, hardwoods, plastics, and composites are used.
8. The substructure consists of posts, beams, ledgers, and joists.
9. See the feature "Plumbing a Post."
10. Sloping allows for drainage.

Contents

Chapter 1	The Construction Industry	191
Chapter 2	Building Codes and Planning	193
Chapter 3	Reading and Drawing Plans	195
Chapter 4	Estimating and Scheduling	197
Chapter 5	Construction Safety and Health	201
Chapter 6	Hand Tools	205
Chapter 7	Power Saws	209
Chapter 8	Electric Drills	211
Chapter 9	Power Tools for Shaping and Joining	215
Chapter 10	Power Nailers and Staplers	219
Chapter 11	Ladders and Scaffolds	221
Chapter 12	Concrete As a Building Material	223
Chapter 13	Locating the House on the Building Site	225
Chapter 14	Foundation Walls	227
Chapter 15	Concrete Flatwork	231
Chapter 16	Wood As a Building Material	233
Chapter 17	Engineered Lumber	235
Chapter 18	Engineered Panel Products	237
Chapter 19	Framing Methods	239
Chapter 20	Floor Framing	241
Chapter 21	Wall Framing and Sheathing	245
Chapter 22	Basic Roof Framing	249
Chapter 23	Hip, Valley, and Jack Rafters	253
Chapter 24	Roof Assembly and Sheathing	257
Chapter 25	Roof Trusses	261

Contents (Continued)

Chapter 26 Steel Framing Basics 263
Chapter 27 Steel Framing Methods 267
Chapter 28 Windows and Skylights 271
Chapter 29 Residential Doors 275
Chapter 30 Roof Coverings 279
Chapter 31 Roof Edge Details 283
Chapter 32 Siding 285
Chapter 33 Brick-Veneer Siding 289
Chapter 34 Stairways 291
Chapter 35 Molding and Trim 295
Chapter 36 Cabinets and Countertops 299
Chapter 37 Wall Paneling 303
Chapter 38 Mechanicals 305
Chapter 39 Thermal and Acoustical Insulation 307
Chapter 40 Walls and Ceilings 309
Chapter 41 Exterior and Interior Finishes 311
Chapter 42 Wood Flooring 313
Chapter 43 Resilient Flooring and Ceramic Tile 315
Chapter 44 Chimneys and Fireplaces 317
Chapter 45 Decks and Porches 319
Answer Key for Chapter Tests 321

The Construction Industry

MATCHING

Directions: Match each word or phrase to its description.

Column I

_____ 1. A worker learning from an expert.

_____ 2. A summary of your experience and qualifications.

_____ 3. Proof of skill in a work area.

_____ 4. Largest group of skilled craftsmen.

_____ 5. Vision, goals, and strategies for a business.

_____ 6. Inner guidelines for telling right from wrong.

_____ 7. Person who creates and runs his or her own business.

_____ 8. System in which businesses can buy, sell, and set prices for goods.

Column II

a. ethics

b. entrepreneur

c. résumé

d. business plan

e. certification

f. apprentice

g. craft

h. carpenters

i. free enterprise

MULTIPLE CHOICE

Directions: In the blank provided, write the letter of the answer that best completes the statement or answers the question.

_____ 9. A recent trend in residential construction in the United States is
   a. multiple subcontractors for a job.
   b. homebuilding boom in warmer climates.
   c. using a variety of foundations.
   d. plumbing and electrical inspections.

_____ 10. Architects would be classified as __?__ workers in the construction industry.
   a. master
   b. craft
   c. technical
   d. professional

_____ 11. Job experience that accompanies a classroom training program is called
   a. an internship.
   b. networking.
   c. an apprenticeship.
   d. entrepreneuring.

(Continued on next page)

CHAPTER 1 TEST (Continued)

_____ 12. What type of job skill is needed to be able to estimate how much lumber you need for a job?
   a. interactive: management
   b. academic: mathematics
   c. academic: scientific
   d. interactive: work ethics

_____ 13. Safety on the job can depend on your ability to read warnings and to make sure that other people _?_ you.
   a. obey
   b. understand
   c. persuade
   d. advise

_____ 14. A friend's younger brother asks you if your company has any summer jobs available. He is gathering information by
   a. mentoring.
   b. thinking critically.
   c. prioritizing.
   d. networking.

_____ 15. If a job application *rightfully* requires information you would rather not tell about yourself, you should
   a. put "NA" for not applicable.
   b. tell the truth.
   c. write a better version than the real one, and explain the details later.
   d. leave the line blank and let them ask why.

_____ 16. The day after an interview,
   a. send the interviewer a small gift or flowers.
   b. send the interviewer a thank-you e-mail.
   c. send the interviewer a thank-you letter.
   d. call the interviewer and thank him or her.

_____ 17. Suppose an employer calls to offer you a job, but you think you might get a better job in a day or two. What should you do?
   a. Ask for time to consider the offer and call back when you promised.
   b. Take the job. You can quit later.
   c. Turn down the job. It may still be open later.
   d. Ask for a very high wage. Either way you win.

_____ 18. If you are hurt on the job and can't work to pay bills, where can you get financial help?
   a. Your employer must loan you money.
   b. Request workers' compensation through your employer.
   c. You may have to rely on family or friends.
   d. Request money to live on from your medical insurance.

_____ 19. In the United States, _?_ laws ensure our right to a minimum wage and protection from discrimination.
   a. business
   b. liability
   c. ethics
   d. labor

_____ 20. Take-home pay, or _?_, is the amount of money you actually receive after deductions.
   a. gross pay
   b. net pay
   c. salary
   d. workers' compensation

Carpentry & Building Construction Instructor Resource Guide
Copyright © Glencoe/McGraw-Hill

Building Codes and Planning

MATCHING

Directions: Match each item to the statement or sentence.

Column I

_____ 1. Establishes minimum standards of quality and safety.

_____ 2. Authorization for the builder to begin construction.

_____ 3. Document indicating that a house is ready to live in.

_____ 4. Minimum distance allowed from a house to a nearby house or the street.

_____ 5. Scale drawing showing the size and location of rooms.

_____ 6. Scale drawing of a standard house plan to show what the finished house will look like.

_____ 7. Specifies such requirements as the minimum-size house for a lot and architectural features.

_____ 8. Long-term loan that is secured by property.

_____ 9. Regulations on which local government can base their own building codes.

_____ 10. Established to make all building codes more uniform.

Column II

a. floor plan
b. deed restrictions
c. setback distance
d. International Residential Code
e. model building code
f. certificate of occupancy
g. construction loan
h. mortgage
i. stock plan
j. building code
k. zoning restrictions
l. building permit

MULTIPLE CHOICE

Directions: In the blank provided, write the letter of the answer that best completes the statement or answers the question.

_____ 11. The three major model building codes used in the United States are the National Building Code, the Standard Building Code, and the
   a. International Residential Code.
   b. National Electrical Code.
   c. American Building Code.
   d. Uniform Building Code.

(Continued on next page)

CHAPTER 2 TEST (Continued)

_____ 12. When planning to build a house, consideration must be given to what the house will look like, how it will fit the family needs, and
   a. whether to obtain a building permit.
   b. how construction will be paid for.
   c. if a spec house should be built first.
   d. whether to use a floor plan or house plan.

_____ 13. What percent of your gross monthly income should be spent for housing?
   a. 80 percent   b. 50 percent   c. 20 percent   d. 25 percent

_____ 14. Besides cost, lot shape, special conditions, and zoning restrictions, what other two factors must be considered when choosing a lot?
   a. location and deed restrictions
   b. model building codes and location
   c. deed restrictions and mortgage payment
   d. certificate of occupancy and location

_____ 15. House plans tailored exactly to the buyers' needs can be obtained from
   a. bankers.   b. inspectors.   c. builders.   d. architects.

_____ 16. Houses in certain neighborhoods may have similar features because the house plans were obtained from the same
   a. banker.   b. inspector.   c. builder.   d. architect.

_____ 17. How are construction costs often estimated?
   a. The loan is obtained and construction costs are considered 80 percent of the total loan.
   b. The average square foot cost for building is multiplied by the total square footage.
   c. The total square footage is divided by 20 percent of the total loan.
   d. The total loan is divided by square foot cost for building.

_____ 18. The loan obtained is commonly what percentage of the total cost?
   a. 80 percent   b. 20 percent   c. 25 percent   d. 100 percent

_____ 19. What legal document is evidence of ownership?
   a. abstract of title   b. contract of sale   c. deed   d. mortgage

_____ 20. After a builder has been chosen, typically a borrower starts with what type of loan from a bank or savings and loan?
   a. mortgage loan   c. home improvement loan
   b. construction loan   d. federal loan

Reading and Drawing Plans

MATCHING

Directions: Match each word or phrase to the statement or sentence.

Column I

_____ 1. Allows the measurements in reduced-scale drawings to be measured in feet and inches.

_____ 2. Imaginary slicing of floor plans.

_____ 3. Numbers that tell the size of a particular feature on a plan.

_____ 4. Site plan, foundation plan, floor plan, and framing plan are several types.

_____ 5. Written notes giving instructions about quality standards for materials and methods of work.

_____ 6. View of an architectural plan that provides information about materials, fastening and support systems, and concealed features.

_____ 7. Shows the exterior of a house as it would look completed.

_____ 8. Sometimes made as isometric drawings to illustrate an assembly detail.

_____ 9. List that contains information about sizes of rough openings, glazing, finish trim and manufacturing name.

_____ 10. Copies of original architectural plans.

Column II

a. schedules
b. architect's scale
c. section view
d. plan view
e. blueprints
f. cutting plane
g. sketch
h. dimensions
i. detail drawing
j. specifications
k. rendering

MULTIPLE CHOICE

Directions: In the blank provided, write the letter of the answer that best completes the statement or answers the question.

_____ 11. The ratio between the size of the objects as drawn and their actual size is called a(n)
    a. dimension.   **c.** scale.
    b. specification.   **d.** elevation.

(Continued on next page)

CHAPTER 3 TEST (Continued)

_____ 12. Which of the following lines used in architectural drawings mark the end points of a dimension and should not touch the outline of the object?
   a. centerlines
   b. extension lines
   c. hidden lines
   d. sectioning lines

_____ 13. Which type of lines on an architectural drawing lead to a part that a note or other reference applies to?
   a. leader lines
   b. break lines
   c. centerlines
   d. hidden lines

_____ 14. Which element of an architectural drawing might clarify the size of, or suggest a construction technique for, an unusually shaped bathtub?
   a. line
   b. symbol
   c. dimension
   d. note

_____ 15. Which of the following statements is true about symbols that are used in architectural drawings?
   a. Symbols describe openings and other changes in exterior walls.
   b. Symbols can be of two kinds: specific and general.
   c. Symbols are used to indicate objects or materials.
   d. Symbols show the shape of an object.

_____ 16. Computer-aided drafting and design programs can be used to create all **except** which of the following?
   a. site plans
   b. sketches
   c. floor plans
   d. elevation drawings

_____ 17. Which of the following is **not** an advantage to using the CADD system?
   a. Drawings can be revised with ease.
   b. CADD software programs are always compatible with each other.
   c. Symbols are easy to add, delete, or move.
   d. A list of materials can be easily produced from the drawing.

_____ 18. The plan view that is relied upon by the excavation contractor is the
   a. mechanical plan.
   b. floor plan.
   c. foundation plan.
   d. site plan.

_____ 19. The drawings in a set of architectural plans that allow the height of objects to be seen are known as
   a. the elevation.
   b. section views.
   c. plan views.
   d. detail drawings.

_____ 20. A list that identifies materials to be used for floors, ceilings, and walls for each room is called a
   a. framing plan.
   b. detail drawing.
   c. room-finish schedule.
   d. specification list.

Estimating and Scheduling

MATCHING

Directions: Match each construction term to its definition or a common example.

Column I

_____ 1. Amount included in an estimate to cover costs of unforeseen situations.

_____ 2. Access charge for phone in builder's truck.

_____ 3. Cost calculation made before the exact features of a house are known.

_____ 4. Dollar figure representing the cost of future choices of products such as cabinetry.

_____ 5. Amount required for materials for building a certain house.

_____ 6. A set of repairs requested by the new house owner before final acceptance.

Column II

a. indirect cost
b. direct cost
c. unit-cost estimate
d. contingency allowance
e. pre-design estimate
f. allowance
g. punch list

(Continued on next page)

Carpentry & Building Construction Instructor Resource Guide
Copyright © Glencoe/McGraw-Hill

CHAPTER 4 TEST (Continued)

MULTIPLE CHOICE

Directions: In the blank provided, write the letter of the answer that best completes the statement or answers the question.

_____ 7. The project scheduling technique that shows the relationships among tasks is the _?_ method.
   a. checklist
   b. critical path
   c. activity coordination
   d. bar charting

_____ 8. A building contractor does a thorough cost estimate for building a house and uses it to prepare a signed proposal called a(n)
   a. conceptual estimate.
   b. CPM.
   c. bid.
   d. allowance.

_____ 9. The construction cost estimate prepared by using average costs per square foot of similar projects is the
   a. quantity takeoff.
   b. unit-cost estimate.
   c. pre-design estimate.
   d. quantity survey.

_____ 10. An estimate that includes specifications such as "22 lineal feet of New Hampshire black granite" is a(n)
   a. unit-cost estimate.
   b. pre-design estimate.
   c. exponential estimate.
   d. quantity takeoff.

_____ 11. An estimate based on the costs of components of a house is a(n)
   a. pre-design estimate.
   b. quantity takeoff.
   c. unit-cost estimate.
   d. quantity survey.

(Continued on next page)

CHAPTER 4 TEST (Continued)

_____ 12. During the construction of a house, the new owner decides she wants Corian countertops instead of plastic laminate countertops. How should the builder handle the request?
   a. Sign and have the owner sign a very specific change order.
   b. Tell the owner how much it will add and write it on the back of the bid sheet.
   c. Very clearly inform the owner of the additional cost and continue work.
   d. Order the Corian countertops and have the bill sent to the owner.

_____ 13. Following are dimensions of four boards. Which one does **not** contain exactly one board foot of lumber?
   a. 2" × 6" × 12"
   b. 1" × 4" × 3'
   c. 1" × 12" × 2'
   d. 2" × 4" × 18"

_____ 14. The Construction Specifications Institute's system of organizing specifications is called
   a. the Standardized Construction Plan.
   b. MasterFormat.
   c. the Construction-Order Checklist.
   d. MasterList.

_____ 15. Utility companies must mark locations of utility lines before
   a. surveying is done.
   b. estimates are prepared.
   c. a building permit can be obtained.
   d. any excavation is done.

_____ 16. All mechanical subcontractors must work in two stages: _?_ work and finish work.
   a. temporary
   b. built-in
   c. rough-in
   d. framing

_____ 17. Besides the amount of labor required, what is a major difference between installing plaster and installing drywall?
   a. Plaster can be installed earlier in the building process.
   b. Drywall must "cure" longer before building can continue.
   c. Plaster takes longer to dry sufficiently before building can continue.
   d. Drywall can be installed earlier in the building process.

(Continued on next page)

Carpentry & Building Construction Instructor Resource Guide
Copyright © Glencoe/McGraw-Hill

CHAPTER 4 TEST (Continued)

_____ 18. Hardwood floors should be finished
   a. after interior painting, carpeting, and cleanup are done.
   b. after interior painting is done and before any carpet is laid.
   c. before any interior finishes like paint and staining are done.
   d. after drywall is installed and before interior painting.

_____ 19. The punch list for a house is prepared by
   a. the owner and the general contractor or builder.
   b. the general contractor or builder.
   c. subcontractors.
   d. the building inspector.

_____ 20. On a CPM diagram, which of these tasks would be shown as being done at the same time?
   a. exterior trim, siding, exterior painting
   b. backfill, sidewalks, landscaping
   c. rough-in electrical, insulation, drywall
   d. finish HVAC, finish electrical, finish plumbing

Construction Safety and Health

MATCHING

Directions: Match each item to the statement or sentence listed below.

Column I

_____ 1. Potential cancer-causing agent found in resins used in construction materials and adhesives.

_____ 2. Cut in the earth made to prepare a site for footings and foundations.

_____ 3. Fast-acting circuit breaker that can protect people from electrical shock.

_____ 4. Combination of temperature and wind speed that increases the chilling point.

_____ 5. Materials that allow electricity to flow through them easily.

_____ 6. Heat-resistant material that causes scarring of the lungs.

_____ 7. Materials that do not allow electricity to flow through them easily.

_____ 8. Provides a path for electricity to flow safely from the tool to the earth.

_____ 9. Electricity flowing through a conductor from a point of origin and back to that same point.

Column II

a. ground-fault circuit interrupter
b. excavation
c. asbestos
d. electrical circuit
e. grounding
f. formaldehyde
g. "hot" wire
h. conductors
i. wind-chill
j. insulators

(Continued on next page)

CHAPTER 5 TEST *(Continued)*

Directions: Match each item to the statement or sentence listed below.

Column I

_____ 10. Tools used to install fasteners into steel, concrete, and masonry.

_____ 11. Caused by lifting, fastening materials, and other tasks.

_____ 12. Irritation of nerves and tissues caused when a task is done over and over.

_____ 13. Tools such as air compressors, chain saws, and portable concrete cutters.

_____ 14. Caused by loss of fluid through sweating when a worker has failed to drink enough fluids.

_____ 15. Occurs when body temperature falls to a level where normal muscle and brain functions are impaired.

_____ 16. Occurs when the body can no longer regulate its core temperature.

_____ 17. Caused by breathing in dust particles produced from cutting concrete, masonry, and rock.

_____ 18. Tools with features such as cushioned grips and properly angled handles.

Column II

a. silicosis
b. ergonomical tools
c. gasoline-powered tools
d. hypothermia
e. powder-actuated tools
f. repetitive stress injury
g. heat exhaustion
h. kickback
i. musculoskeletal disorder
j. heat stroke

MULTIPLE CHOICE

Directions: In the blank provided, write the letter of the answer that best completes the statement or answers the question.

_____ 19. The purpose of OSHA is to assure, as far as possible, safe and healthful working conditions and to
  a. underwrite workers' compensation programs.
  b. provide material safety data sheets (MSDS).
  c. research and produce ergonomically designed tools.
  d. preserve human resources.

_____ 20. When practicing good housekeeping on the job,
  a. use caution signs to mark walkways cluttered with tools and materials.
  b. tag danger signs on protruding nails.
  c. dispose of scraps and rubbish daily to prevent fires.
  d. barricade unsanitary toilets.

(Continued on next page)

Name _____ Date _____ Class _____

CHAPTER 5 TEST (Continued)

_____ 21. When using colors on signs, tags, and barricades, red means danger, yellow means caution, and orange means
  a. first aid.
  b. be on guard.
  c. storage.
  d. information.

_____ 22. Kickback is a hazard occurring when operating
  a. power saws.
  b. portable tools with abrasive wheels.
  c. powder-actuated tools.
  d. gasoline-powered tools.

_____ 23. A hazard of using gasoline-powered tools is that
  a. cartridges can explode unexpectedly.
  b. flying fragments may be thrown off.
  c. fuel vapors can burn or explode.
  d. they may disintegrate during start up.

_____ 24. To warn workers about an open stairwell in a construction site, a ? sign should be used.
  a. caution (yellow)
  b. danger (red)
  c. be on guard (orange)
  d. lockout

_____ 25. To prevent heat-related health problems, you should
  a. eat plenty of food.
  b. drink plenty of water.
  c. be careful around excavations.
  d. use quality work practices.

_____ 26. The four factors that cause cold-related stress are low temperatures, high/cool winds, cold water, and
  a. diet.
  b. humidity.
  c. sunlight.
  d. dampness.

_____ 27. Which of the following can help protect a worker from repetitive stress injury?
  a. pneumatic nailers and knee pads
  b. portable saw horses
  c. safety harnesses
  d. ground fault circuit interrupters

(Continued on next page)

CHAPTER 5 TEST (Continued)

_____ 28. Which of the following statements is true about the need for personal protective devices?
   a. Hearing protectors will eliminate all noise.
   b. OSHA requires the use of hard hats on all job sites.
   c. Eye protection should be worn on all job sites.
   d. A dust mask should be worn to protect against fumes or fine dust.

_____ 29. Slightly high body temperature, extreme weakness, nausea, or headache are primarily symptoms of
   a. hypothermia.
   b. heat exhaustion.
   c. heat stroke.
   d. heat cramps.

_____ 30. The most common causes of injury on a construction site are
   a. fires.
   b. injuries from falling objects.
   c. falls.
   d. faulty equipment.

_____ 31. Wearing a hat, sunglasses, and sunscreen are precautions that should be taken to protect workers from
   a. cold.
   b. heat.
   c. Lyme disease.
   d. sunlight.

_____ 32. Information explaining the hazards associated with certain dangerous substances can be found in
   a. Right to Know (RTK) laws.
   b. Material Safety Data Sheets (MSDS).
   c. American National Standards Institute (ANSI) regulations.
   d. National Institute for Occupational Safety and Health (NIOSH) regulations.

_____ 33. Many builders feel that there is a direct relationship between quality work practices and safety because
   a. the concentration level is worse when the work is constant.
   b. workers are less likely to be distracted if they rush to complete jobs.
   c. workers who take a long time to do a job have a greater chance for errors.
   d. workers who have a high degree of craftsmanship are safer workers.

Hand Tools

MATCHING

Directions: Match each item shown in Fig. 6-1 to the correct tool.

_____ **1.** try square

_____ **2.** sliding T-bevel

_____ **3.** backsaw

_____ **4.** keyhole saw

_____ **5.** coping saw

Fig. 6-1

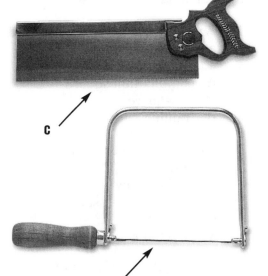

(Continued on next page)

CHAPTER 6 TEST (Continued)

Directions: Match each tool with its use.

Column I

_____ 6. Can be used to measure a walkway or driveway.

_____ 7. Used to create a quick cutting line over long lengths.

_____ 8. A pointed metal marking tool used to start a hole before drilling.

_____ 9. Used to fit doors and plane end grain.

_____ 10. Used to plane long pieces such as the edges of doors.

Column II

a. block plane
b. chalk line
c. jointer plane
d. scratch awl
e. layout tape measure

MULTIPLE CHOICE

Directions: In the blank provided, write the letter of the answer that best completes the statement or answers the question.

_____ 11. Which of the following is a blade that slides along a handle and may contain a leveling vial and a scriber?
   a. sliding T-bevel
   b. combination square
   c. try square
   d. framing square

_____ 12. When cutting curves and enlarging holes, which of the following saws should be used?
   a. keyhole saw
   b. utility drywall saw
   c. hacksaw
   d. dovetail saw

_____ 13. When shaping the end of molding for joints, the best saw to use would be a(n)
   a. utility drywall saw.
   b. dovetail saw.
   c. coping saw.
   d. keyhole saw.

(Continued on next page)

Name _____ Date _____ Class _____

CHAPTER 6 TEST (Continued)

_____ 14. Which of the following saws is used to cut molding and trim and is often used in a miter box?
   a. handsaw
   b. ripsaw
   c. dovetail saw
   d. backsaw

_____ 15. Which of the following types of hammers can be used for prying apart pieces that have been nailed together as well as for driving and removing nails?
   a. rip hammer
   b. claw hammer
   c. Warrington hammer
   d. hand sledge

_____ 16. A tool that will substitute for an adjustable wrench, vise, clamp, or pipe wrench is
   a. slip-joint pliers.
   b. locking pliers.
   c. box-joint utility pliers.
   d. needle-nose pliers.

_____ 17. Another name for groove-joint pliers or waterpump pliers is
   a. Lineman's pliers.
   b. slip-joint pliers.
   c. locking pliers.
   d. box-joint utility pliers.

_____ 18. Which of the following wrenches is used to tighten or loosen set screws and is sometimes called a hex key?
   a. adjustable wrench
   b. open-end wrench
   c. box wrench
   d. Allen wrench

(Continued on next page)

CHAPTER 6 TEST (Continued)

_____ 19. Another name for a tool sometimes called side cutters and is used to cut electrical and other wire is
   a. tin snips.
   b. aviation snips.
   c. Lineman's pliers.
   d. needlenose pliers.

_____ 20. A saw very much like a backsaw used to make cuts of superior accuracy is a
   a. hacksaw.
   b. dovetail saw.
   c. coping saw.
   d. keyhole saw.

Power Saws

MATCHING

Directions: Match the type of power-saw cut to its description listed below.

Column I

_____ 1. Made at an angle across the grain of the material.

_____ 2. Made across the direction of the material's grain.

_____ 3. Made at an angle through the thickness of the material.

_____ 4. Does not cut through to an edge of the material.

_____ 5. Made parallel to the grain of the material.

Column II

a. bevel cut
b. miter cut
c. rip cut
d. offcut
e. internal cut
f. cross cut

Directions: Match each cutting material or need to the best saw/blade combination for that purpose.

Column I

_____ 6. Circular saw/circular blade with perimeter teeth.

_____ 7. Miter saw/circular blade with 60+ teeth.

_____ 8. Circular saw/circular blade with abrasive.

_____ 9. Reciprocating saw/straight blade, bimetal.

_____ 10. Circular saw/circular blade with grinding edge, dry.

Column II

a. ceramic tile
b. wood or metal sheets
c. fine cuts on expensive molding
d. demolition having old nails
e. brick or stone
f. solid metal

(Continued on next page)

CHAPTER 7 TEST (Continued)

MULTIPLE CHOICE

Directions: In the blank provided, write the letter of the answer that best completes the statement or answers the question.

_____ 11. Worm-drive saws require a special _?_ for the gears.
   a. lubricant
   b. baseplate
   c. blade guard
   d. power cord

_____ 12. If the blade of a circular saw pinches or binds, what reaction can occur?
   a. The lade teeth may fly off.
   b. The gears will burn out.
   c. The saw may kick back.
   d. The saw motor will die.

_____ 13. When cutting cabinet boards with a circular saw, how should you cut them to get the smoothest cut?
   a. face up
   b. face down
   c. as slowly as possible
   d. with the grain

_____ 14. The primary use of a table saw is for
   a. creating beveled edges.
   b. ripping stock to width.
   c. cutting joining sections.
   d. performing demolition jobs.

_____ 15. The part of a table saw that guides stock as it is cut is called a
   a. splitter.
   b. throat plate.
   c. rip fence.
   d. blade guard.

_____ 16. A table saw that is on a fixed table and has a blade that can be moved and angled is a
   a. splitter saw.
   b. tilting-arbor saw.
   c. mitered-arbor saw.
   d. fixed-arbor saw.

_____ 17. You should always use a(n) _?_ to finish ripping narrow stock on a table saw.
   a. assistant
   b. blade-angle wheel
   c. miter gauge
   d. push stick

_____ 18. You should always use hearing protection when using power saws, especially the whining
   a. reciprocating saw.
   b. jigsaw.
   c. miter saw.
   d. table saw.

_____ 19. When you're cutting with a jigsaw, the "good" surface of the material should face
   a. at a 45° angle.
   b. at a 90° angle.
   c. down.
   d. up.

_____ 20. As with any power saw, the blade of a jigsaw should be at _?_ when you move it into the material.
   a. low speed
   b. introductory speed
   c. full speed
   d. medium speed

Electric Drills

MATCHING

Directions: Match each word or phrase to the electric drill safety guideline that it best completes.

Column I

_____ 1. When drilling with an electric drill, use steady pressure but don't _?_ the drill.

_____ 2. Because cloth will _?_ around a drill bit, you should not drill through it.

_____ 3. Before installing a bit into an electric drill, you should _?_ the power source.

_____ 4. Never use a bit with a square, tapered tang in an electric drill because the chuck cannot _?_ it securely.

_____ 5. When drilling holes in small parts, always _?_ the parts to prevent them from spinning.

_____ 6. Before you start to use an electric drill, _?_ the chuck key from the drill.

Column II

a. disconnect
b. remove
c. clamp
d. hold
e. force
f. twist

Directions: Match each type of electric drill bit with its common usage.

Column I

_____ 7. In wood, it bores smooth holes with flat bottoms.

_____ 8. Used on brick, concrete, and other masonry materials.

_____ 9. Creates a funnel shape at the top of a drilled hole.

_____ 10. Cuts deep holes quickly in wood; often used by electricians.

_____ 11. General utility bit for drilling many materials.

_____ 12. Cuts an exceptionally smooth, clean hole in wood.

Column II

a. brad point bit
b. twist bit
c. masonry bit
d. auger bit
e. countersink bit
f. forstner bit

(Continued on next page)

CHAPTER 8 TEST *(Continued)*

MULTIPLE CHOICE

Directions: In the blank provided, write the letter of the answer that best completes the statement or answers the question.

_____ 13. Why would you want your electric drill with a metal housing to have a three-prong plug?
   a. to ensure that it gets full power input
   b. to reduce the chances of getting shocked
   c. because you get cleaner power, ensuring higher torque
   d. because you can't plug two-prong plugs into most outlets

_____ 14. The type of electric drill best suited for drilling large holes or holes in steel or concrete is a
   a. cordless drill.
   b. right-angle drill.
   c. corded drill.
   d. drill-driver.

_____ 15. What number indicates the power of a battery-powered drill?
   a. the voltage of the battery
   b. the drill's generic model number
   c. the total number of batteries it uses
   d. all battery-powered drills are 12 volts

_____ 16. The best electric drill for a worksite located half a mile from roads and utilities would be
   a. a high-torque, corded drill.
   b. a high-rpm drill.
   c. a high-amperage drill.
   d. a cordless drill.

_____ 17. Where can you find the amperage rating of an electric drill?
   a. in the drill's Owner's Manual
   b. on a plate on the drill
   c. on a tag attached to the drill
   d. stamped inside the drill housing

(Continued on next page)

Name _____ Date _____ Class _____

CHAPTER 8 TEST *(Continued)*

_____ 18. Which of the following describes the correct drilling procedure?
   a. Apply as much pressure as possible; move drill gently from side to side.
   b. Press drill very firmly to start, then rotate drill in a circular motion.
   c. Use a sharp bit; start by drilling at a right angle.
   d. Drill the hole completely, stop the drill, and remove the bit from wood.

_____ 19. Why is a drill bit hot after drilling into a piece of wood?
   a. from the electricity going through it
   b. from the friction against the wood
   c. from the chemicals contained in the wood
   d. from being near the motor of the drill

_____ 20. When you're drilling hardwood, a pilot hole should be
   a. one size smaller than the root diameter of the screw.
   b. two sizes smaller than the root diameter of the screw.
   c. one size larger than the root diameter of the screw.
   d. the same size as the root diameter of the screw.

_____ 21. What situation does **not** require the creation of a pilot hole?
   a. when putting drywall screws into softwoods
   b. when using fully threaded screws
   c. when putting drywall screws into hardwoods
   d. when fastening two pieces of wood together

_____ 22. What type of bit is best for drilling into metals?
   a. a hardened metal bit or masonry bit
   b. a twist bit with a 135° split point
   c. a twist bit with a 90° split point
   d. a screwdriving bit

_____ 23. If you're drilling into stainless steel and want to extend the life of a drill bit, what can you do?
   a. Run water over the surface as you drill.
   b. Buff down the metal surface first to make drilling easier.
   c. Use a cutting lubricant to cool the bit.
   d. Heat the metal slightly to make it softer.

(Continued on next page)

CHAPTER 8 TEST (Continued)

_____ **24.** You can easily check the straightness of a twist bit by
   a. spinning it in the drill and watching for wobbling.
   b. drilling through a tube-shaped piece of test material.
   c. checking it visually from the side.
   d. checking it visually from the top.

_____ **25.** If you are using a drill to drive a screw into an expensive piece of wood and can't risk causing damage, what would be the best method to use?
   a. Start with a screwdriver and finish with an electric drill.
   b. Drill quickly and finish without stopping the drill.
   c. Start slowly and speed up after it's halfway through.
   d. Drive it partway with an electric drill and finish with a screwdriver.

Power Tools for Shaping and Joining

MATCHING

Directions: Match each tool to the safety tip listed below that should go with it.

Column I

_____ 1. Keep knives sharp to avoid kickback.

_____ 2. Make multiple shallow cuts in hardwoods.

_____ 3. Bring to full speed before cutting.

_____ 4. Do not use to remove paint with lead.

_____ 5. Ensure proper tracking on rollers.

_____ 6. Keep face away from dust and chips that are ejected quickly.

Column II

a. all sanders
b. all routers
c. jointer
d. routing biscuit
e. plate joiner
f. belt sander
g. planer

Directions: Match each power tool to its primary use from the list below.

Column I

_____ 7. Trims and installs paneling.

_____ 8. Creates right angles on cabinets.

_____ 9. Squares up stair balusters.

_____ 10. Cuts a pilot hole.

_____ 11. Does final wood finishing.

_____ 12. Trims plastic laminate.

Column II

a. router
b. plunge router
c. jointer
d. planer
e. power plane
f. flush-trimming router
g. orbital sander

(Continued on next page)

CHAPTER 9 TEST (Continued)

MULTIPLE CHOICE

Directions: In the blank provided, write the letter of the answer that best completes the statement or answers the question.

_____ 13. What tool would you use to make a half-round edge of a countertop?
  a. jointer
  b. belt sander
  c. router
  d. planer

_____ 14. A router's size is usually classified by its
  a. motor type.
  b. standard application.
  c. horsepower.
  d. collet.

_____ 15. What can be added to a router bit to change the profile or depth of the cut?
  a. bearing
  b. second bit
  c. bit guard
  d. shank

_____ 16. What router template would help you create interlocking drawer pieces?
  a. edge guide
  b. plunge
  c. dovetail
  d. mortise

_____ 17. What is the best way to make deep cuts with a router?
  a. with the longest bit available
  b. in several passes
  c. using very firm pressure
  d. on a router table

_____ 18. A belt sander's size is classified by
  a. horsepower.
  b. its speed.
  c. belt length and width.
  d. typical applications.

_____ 19. When you install a belt on a portable belt sander, what do you align with the arrow on the sander?
  a. the belt's lap seam
  b. the lock on top of sander
  c. the top handle of the sander
  d. the arrow inside the belt

(Continued on next page)

Name _____ Date _____ Class _____

CHAPTER 9 TEST (Continued)

_____ 20. How should you operate a belt sander to keep from gouging the wood?
   a. Keep it moving.
   b. Operate at slow speeds.
   c. Apply only the front half of belt.
   d. Press lightly, using one hand.

_____ 21. An orbital sander's size is determined by
   a. belt width.
   b. size of pad.
   c. amps.
   d. abrasiveness.

_____ 22. What type of sander can be used either with the wood's grain or against it?
   a. 4" by 24" belt sander
   b. finishing sander
   c. orbital sander
   d. random-orbit sander

_____ 23. An 8-inch jointer is a common size. What does the size measure?
   a. maximum possible width of cut
   b. minimum blade size
   c. maximum angle of joint created
   d. minimum possible width of cut

_____ 24. A planer's size is based on the size of its ? and therefore the width of board it can plane.
   a. blade
   b. bed
   c. infeed rollers
   d. handwheel

_____ 25. How should stock be fed into a planer?
   a. The infeed roller should be allowed to pull it through.
   b. The operator should push it through with moderate force.
   c. The outfeed roller should pull it through.
   d. It should be pushed through with a push stick.

_____ 26. What tool would be most useful if you needed to trim and square paneling at a worksite?
   a. plate joiner
   b. portable jointer
   c. plunge router
   d. portable electric plane

(Continued on next page)

CHAPTER 9 TEST (Continued)

_____ 27. Why is a plate joiner sometimes called a biscuit joiner?
   a. It cuts edge grooves into which small wood pieces called biscuits are glued.
   b. It can cut three-dimensional circular shapes that look like biscuits.
   c. It cuts smooth round holes in the center of materials.
   d. The cutting head looks like a biscuit cutter.

_____ 28. A plate joiner has _?_ points on its faceplate that should never be disabled.
   a. cutting fence
   b. sawdust deflection
   c. anti-kickback
   d. guide

_____ 29. Any portable power finishing equipment should be temporarily _?_ during use.
   a. operated at low power
   b. switched to AC power
   c. disconnected from a vacuum system
   d. secured to floor or workbench

_____ 30. Always make sure a power tool is _?_ when you install, remove, or adjust bits.
   a. turned off
   b. set down firmly
   c. pointing away from you
   d. unplugged from its power source

_____ 31. When using a jointer, always stand
   a. behind it.
   b. to the side of it.
   c. far away from it.
   d. in front of it.

_____ 32. A router is a portable tool used primarily for
   a. smoothing the surface boards
   b. drilling large holes in finished boards.
   c. finishing work such as shaping edges.
   d. removing heavy stock before planing.

_____ 33. All planers have a _?_ that moves the bed up and down to control the depth of the cut.
   a. press-button switch
   b. handwheel
   c. set of adjustable screws
   d. semi-automatic lever

Power Nailers and Staplers

MATCHING

Directions: Match each type of fastener to its most common carpentry application, listed below.

Column I

_____ 1. Fastening trims.

_____ 2. Nailing hardwoods.

_____ 3. Nailing subfloors.

_____ 4. Framing and general construction.

_____ 5. Attaching roofing.

_____ 6. Fastening softwoods.

Column II

a. smooth-shank nails
b. divergent-point staples
c. screw-shank nails
d. ring-shank nails
e. divergent-point nails
f. fine-wire, narrow crown staples
g. heavy-wire, wide crown staples

Directions: Match each part of a pneumatic fastening system to its function, listed below.

Column I

_____ 7. Contains piston and driver blades.

_____ 8. Transfers compressed air to tool.

_____ 9. Air-driven nailer.

_____ 10. Source of air.

_____ 11. Presses against workpiece.

Column II

a. pneumatic tool
b. high-pressure hose
c. compressor
d. tool head
e. wide-crown staples
f. nosepiece

(Continued on next page)

Carpentry & Building Construction Instructor Resource Guide
Copyright © Glencoe/McGraw-Hill

CHAPTER 10 TEST *(Continued)*

MULTIPLE CHOICE

Directions: In the blank provided, write the letter of the answer that best completes the statement or answers the question.

_____ 12. The two-step firing sequence of newer power fasteners requires that the tool's nosepiece be _?_ or the tool won't fire.
   a. pointed downward
   b. double-triggered
   c. against the workpiece
   d. perfectly clean

_____ 13. A cordless nailer is powered by a(n)
   a. electrical generator.
   b. rechargeable battery.
   c. air compressor.
   d. combustion engine.

_____ 14. An air hose for a power nailer should be rated at _?_ the maximum output of the compressor.
   a. the same pressure as
   b. 50% higher than
   c. lower than
   d. three times

_____ 15. Which of these air hose sizes would be correctly sized for a power nailer?
   a. 5/16" × 125'
   b. 5/16" × 100'
   c. 1/2" × 50'
   d. 1/4" × 50'

_____ 16. The crown of a staple that was power-driven correctly is
   a. parallel to the wood grain.
   b. just slightly above the wood's surface.
   c. perpendicular to the wood grain.
   d. diagonally across the wood grain.

_____ 17. What type of device holds power-driven fasteners in a loop arrangement?
   a. coil-loaded magazine
   b. strip-loaded magazine
   c. hoop magazine
   d. straight magazine

_____ 18. Nails with D-shaped heads may not hold as well as round-head nails, but they
   a. always meet code.
   b. are cheaper.
   c. are easier to drive.
   d. are packed tightly.

_____ 19. Power fasteners can be held together in loops or strips by means of
   a. paper or plastic bands.
   b. glue.
   c. high pressure.
   d. rubber bands.

_____ 20. You should release the pressure from an air compressor at the end of every work day and
   a. before transporting it.
   b. when it's unused for over an hour.
   c. during your breaks.
   d. if it's used indoors.

Ladders and Scaffolds

MATCHING

Directions: Match each item to the statement or sentence listed below.

Column I

_____ 1. Type of ladder that is self-supporting and is used primarily indoors for low and intermediate heights.

_____ 2. Type of ladder used primarily outdoors where greater heights must be reached.

_____ 3. Holds the stepladder open and prevents it from accidentally closing.

_____ 4. A portable metal frame designed specifically to support scaffolding.

_____ 5. Metal devices that slide up and down on a wood or aluminum post to support horizontal planks used on a scaffold.

_____ 6. A rope intended to prevent a worker from falling more than 6 feet.

_____ 7. Rungs or steps attached to vertical supports.

_____ 8. Can be attached to the frame of a structure to allow scaffold planks to form a platform.

_____ 9. A common type of folding ladder that has flattened steps instead of rungs.

_____ 10. Bolted to the top of an extension ladder to steady the top and prevent slipping.

_____ 11. A raised platform that is used for working at heights safely with both hands free.

_____ 12. Horizontal parts of a scaffold used because of their quality and extra strength.

_____ 13. Straight ladder that can be adjusted to various lengths.

Column II

a. trestle
b. brackets
c. ladder stabilizer
d. laminated wood planks
e. folding
f. rails
g. straight
h. scaffold
i. spreader
j. extension ladder
k. articulated
l. lifeline
m. stepladder
n. pump jacks

(Continued on next page)

CHAPTER 11 TEST (Continued)

MULTIPLE CHOICE

Directions: In the blank provided, write the letter of the answer that best completes the statement or answers the question.

_____ 14. What type of ladder can be adjusted to fit into a stairwell?
   a. stepladder
   b. articulated ladder
   c. straight ladder
   d. extension ladder

_____ 15. The first step when setting up a straight ladder is to
   a. grasp a rung at the upper end with both hands.
   b. raise the top end and carefully walk forward.
   c. brace the lower end against a step or other object.
   d. check the angle, height, and stability at the top and bottom.

_____ 16. Which of the following support equipment do contractors sometimes use for working on ceilings?
   a. trestles
   b. sawhorses
   c. pump jacks
   d. corner brackets

_____ 17. Which of the following support equipment is commonly used to reach the side walls of a house during siding or painting operations?
   a. trestles
   b. pump jacks
   c. roofing brackets
   d. corner brackets

_____ 18. Which of these is an advantage that metal scaffolding has over wood scaffolding?
   a. It can be rented as needed.
   b. It is fixed in place at all times.
   c. It is heavier in weight than wood scaffolding.
   d. It takes up less space than wood scaffolding.

_____ 19. Which of the following is a general ladder safety rule?
   a. Never place a ladder in front of a downspout.
   b. Always inspect ladders and keep nuts and bolts tightened.
   c. Never use a laminated wood plank as a scaffolding plank.
   d. Always make sure a stepladder is open and the spreader is unlocked.

_____ 20. When using a stepladder, make sure
   a. the ladder stabilizer can be bolted to the top of the ladder.
   b. the top step extends above the edge of a roof by three inches.
   c. the spreader is locked into place.
   d. tools are always placed on the top step.

Concrete As a Building Material

MATCHING

Directions: Match each concrete-related term to its definition listed below.

Column I

_____ 1. Any unusual ingredient added to a concrete mix to change its characteristics.

_____ 2. The granular component of concrete, such as sand or gravel.

_____ 3. The active component of concrete that is made of lime, silica, and alumina.

_____ 4. The chemical reaction that occurs when concrete is mixed.

_____ 5. An ingredient added to concrete to allow it to be put into use quickly.

Column II

a. hydration
b. aggregate
c. Portland cement
d. admixture
e. ASTM
f. accelerator

Directions: Match each type of cement to its definition listed below.

Column I

_____ 6. Contains fine aggregate and can be spread in thin layers.

_____ 7. Standard grade; used for most general concrete construction.

_____ 8. High-strength; used where concrete forms must be removed quickly.

_____ 9. Expands when mixed with water; used to plug cracks in foundations.

_____ 10. Prevents cracking caused by temperature changes; used for large projects like dams.

Column II

a. Type I
b. Type VI
c. Type III
d. Type IV (low heat)
e. resurfacing
f. hydraulic

(Continued on next page)

CHAPTER 12 TEST (Continued)

MULTIPLE CHOICE

Directions: In the blank provided, write the letter of the answer that best completes the statement or answers the question.

_____ 11. What characteristic of concrete is *most* important to its use in foundations?
  a. highly resistant to chemicals
  b. can be formed into almost any shape
  c. high compressive strength
  d. widely available

_____ 12. As more ? is added to cement, its compressive and tensile strength decrease.
  a. sand   **b.** gravel   **c.** lime   **d.** water

_____ 13. What should you do to new concrete as it cures to improve its strength?
  a. Eliminate all movement near it.
  b. Coat it with waterproof sealants immediately.
  c. Protect it from sunlight's ultraviolet rays.
  d. Keep it moist.

_____ 14. Which of the following is **not** a characteristic of concrete?
  a. It is impervious to rot.
  b. Insects don't damage it.
  c. It is a relatively expensive building material.
  d. It can withstand extreme heat or cold.

_____ 15. Because concrete can be ? easily, it is possible to use it to make countertops.
  a. cured   **b.** hydrated   **c.** formed   **d.** mixed

_____ 16. What quality of water is needed for a concrete mix?
  a. clean enough to drink
  b. reasonably clear
  c. spring
  d. distilled

_____ 17. How should water be added to concrete being mixed by hand?
  a. Make a well in center of dry mix; put water into well; stir in a circle.
  b. Mix in half the water, then very gradually add the rest, checking consistency.
  c. Mix in ¼ of the water at a time; let concrete stand 15 minutes between additions.
  d. Have one person mix continuously as someone else adds the intended amount of water.

_____ 18. If you open a sack of pre-mixed concrete and it has very hard lumps in it, what should you do?
  a. Be sure to pulverize all lumps before mixing with water.
  b. Don't use it; it has gotten wet somehow.
  c. Adjust the amount of water you use; it will likely take more.
  d. Add more lime to the mixture to make up for the lumps.

_____ 19. What would a 1:2:3 batch of good concrete mix, measured in bucketfuls, have in it?
  a. 1 bucket Portland cement, 2 buckets sand, 3 buckets gravel
  b. 1 bucket gravel, 2 buckets Portland cement, 3 buckets sand
  c. 1 bucket sand, 2 buckets gravel, 3 buckets Portland cement
  d. 1 bucket Portland cement, 2 buckets water, 3 buckets sand/gravel mix

_____ 20. Ready-mix concrete is the most economical choice if you need ? or more.
  a. 2 cubic feet   **b.** 4 cubic feet   **c.** 2 cubic yards   **d.** 4 cubic yards

Locating the House on the Building Site

MATCHING

Directions: Match each item to the statement or sentence listed.

Column I

_____ 1. Used to locate and mark the outline of the building.

_____ 2. Any point of reference from which measurements can be made.

_____ 3. Shows the location of the building on the lot, along with related land elevations.

_____ 4. Point over which the level is directly centered and from which the layout is sighted.

_____ 5. Transit that reads horizontal and vertical angles electronically.

_____ 6. Transit that reads horizontal and vertical angles on a vernier scale.

_____ 7. Surveying instruments that receive signals from satellites orbiting the earth.

_____ 8. Surveying instruments that use an electronic detector to sense focused beams of light.

_____ 9. Makes it possible to stake out the site without using a surveying instrument.

_____ 10. Minimum distance allowed by local codes between a house and the property lines.

Column II

a. theodolite
b. leveling rod
c. batter board
d. setback distance
e. GPS instruments
f. bench mark
g. reference line
h. laser instruments
i. plot plan
j. transit level
k. station mark

MULTIPLE CHOICE

Directions: In the blank provided, write the letter of the answer that best completes the statement or answers the question.

_____ 11. To stake out a simple building layout without using a surveying instrument, use a guide such as a
   a. setback distance.
   b. reference line.
   c. plot plan.
   d. global positioning system.

(Continued on next page)

CHAPTER 13 TEST (Continued)

_____ 12. When using existing lines such as a street or marked property line to stake out a site, the first step would be to
   a. check the plot plan to determine the depth of the building.
   b. establish the rear corners of the building by laying out a right angle.
   c. check the plot plan to find the setback distance.
   d. find out how far the corners should be from the side boundaries.

_____ 13. The least expensive surveying instrument is the
   a. laser level.
   b. transit level.
   c. builder's level.
   d. automatic level.

_____ 14. Which of the following surveying instruments does not use a telescope?
   a. laser level
   b. transit level
   c. builder's level
   d. automatic level

_____ 15. Which of the following steps is correct when laying out a right angle using a builder's level?
   a. Use a plumb bob to set up the level directly over the reference point.
   b. Sight the point where the right angle is to be located.
   c. Set up the transit or level directly over the point where the right angle is to be located.
   d. Once a reference point has been sighted, set the 360-degree scale at 90 degrees.

_____ 16. The height of batter boards is sometimes the height of the
   a. foundation wall.
   b. footings.
   c. first floor.
   d. grade level.

_____ 17. Using 12 feet and 16 feet for the legs of the corner in the 3-4-5 method, what is the diagonal length?
   a. 18 feet
   b. 21 feet
   c. 24 feet
   d. 20 feet

_____ 18. To measure a difference in elevation between two points, set up the transit
   a. at right angles to the first point.
   b. at the second point.
   c. on the first point.
   d. at an intermediate point.

_____ 19. Estimate the volume of soil excavated for a house foundation measuring 44 feet by 28 feet by 7 feet deep.
   a. 398 cubic yards
   b. 319 cubic yards
   c. 358 cubic yards
   d. 411 cubic yards

_____ 20. Which of the following can be used as a reference line when staking out a building site?
   a. a batter board line
   b. a street
   c. a section line
   d. a sewer line

Foundation Walls

MATCHING

Directions: Match each term to its definition listed below.

Column I

_____ 1. Horizontal bracing member in reusable concrete forms.

_____ 2. Vertical mortar junctions in masonry.

_____ 3. Masonry joints that permit slight movement.

_____ 4. Coating of mortar or cement applied to block walls for moisture-proofing.

_____ 5. Horizontal mortar joints used in concrete masonry.

_____ 6. Occurs when fresh concrete is poured on top of cured concrete.

Column II

a. cold joint
b. parging
c. head joint
d. control joint
e. rebar
f. bed joint
g. wale

Directions: Match each item listed below to its use.

Column I

_____ 7. Embedded in wet concrete to improve footing strength.

_____ 8. Contain the wet concrete when it is poured.

_____ 9. A larger bearing surface at a structure's base.

_____ 10. Support floor girders in crawl-space homes.

_____ 11. Left in place permanently as part of the structure.

Column II

a. footings
b. keyways
c. reinforcing bar
d. piers
e. forms
f. insulating concrete forms (ICF)

(Continued on next page)

CHAPTER 14 TEST (Continued)

MULTIPLE CHOICE

Directions: In the blank provided, write the letter of the answer that best completes the statement or answers the question.

_____ 12. Depending on the location, footings may be placed more than 12" into the ground in order to
   a. keep pests such as termites out of the structure.
   b. prevent frost damage to the foundation.
   c. create below-grade water drainage channels.
   d. create a better external appearance.

_____ 13. On standard soil, a footing's width should be
   a. as wide as the foundation wall.
   b. one-half times as wide as the foundation wall.
   c. twice as wide as the foundation wall.
   d. three times as wide as the foundation wall.

_____ 14. What type of footings would be needed on soil with low load-bearing capacity?
   a. specially formulated concrete footings
   b. concrete masonry block footings
   c. concrete spread footings
   d. wider and thicker footings

_____ 15. Where are stepped footings used?
   a. where the building plan indicates steps will be built
   b. on building sites where the lot slopes
   c. under the multi-story structure foundations
   d. in all earthquake-prone locations

_____ 16. Building codes generally require a foundation that encloses usable space below grade to have
   a. foundation drains.
   b. waterproof insulation.
   c. metal termite barriers.
   d. sump pumps.

_____ 17. Foundation, or footing, drains are required where
   a. storm sewers are installed.
   b. a building lot is above grade.
   c. heavy runoff occurs.
   d. poured concrete meets soil.

_____ 18. Form-grade plywood made by APA certified mills is called
   a. Millform.
   b. Surform.
   c. Formstay.
   d. Plyform.

(Continued on next page)

CHAPTER 14 TEST (Continued)

_____ 19. How long should standard concrete wall forms be left in place to enhance curing?
   a. 12 hours
   b. 12 to 24 hours
   c. 36 hours
   d. 3 to 7 days

_____ 20. Which of the following is **not** true of insulating concrete forms?
   a. They are made of rigid foam insulation.
   b. The foam is removed after the concrete cures.
   c. They are light and easy to install.
   d. They can be used both below and above grade.

_____ 21. In order to avoid creating cold joints in concrete, it should be poured
   a. continuously, without interruption.
   b. only into prepared plywood forms.
   c. only when the air temperature is above freezing.
   d. in intersecting sections at different times.

_____ 22. A horizontal support member used above an opening in a concrete foundation wall is called a
   a. sill plate.
   b. veneer.
   c. lintel.
   d. header.

_____ 23. What would you put into a foundation to prepare for future plumbing or electrical components?
   a. anchor bolts
   b. foundation vents
   c. utility sleeves
   d. metal conduits

_____ 24. Why is it a good idea to build the floor on top of a foundation before backfilling?
   a. It is easier to access the floor supports.
   b. It ensures that foundation walls hold up against the backfill weight.
   c. Building inspectors will want to see the outside of the foundation.
   d. Foundation waterproofing can be tested during a rain.

_____ 25. When are anchor bolts installed in a foundation?
   a. as the first floor decking is being added
   b. when the foundation concrete is wet
   c. when the footing concrete is still wet
   d. after the first floor framing is finished

(Continued on next page)

CHAPTER 14 TEST (Continued)

_____ 26. Before backfilling is done, any foundation should be treated like
   a. any other non-waterproof wall.
   b. concrete that is already cured.
   c. a potential safety hazard.
   d. an unsupported footing wall.

_____ 27. One main advantage of using concrete block versus poured concrete for foundations is
   a. forms for concrete block are simpler to use.
   b. concrete blocks are light and easy to handle.
   c. mortar does not have to be protected from cold.
   d. work can be stopped and started if necessary.

_____ 28. If a very strong mortar mixture is needed, what component should be increased in the formula?
   a. contractor's sand
   b. Portland cement
   c. hydrated lime
   d. pulverized gravel

_____ 29. After mortar is mixed, it should be used within
   a. one hour.
   b. 1 to 2 hours.
   c. 2 ½ and 3 ½ hours.
   d. six hours if you keep adding water.

_____ 30. How should you begin constructing a block wall?
   a. start at the end that has a door
   b. build from the middle and work out
   c. lay one complete course of block the length of the wall
   d. build the corners first, several courses high

_____ 31. A _?_ joint should be used at the intersection of bearing walls made of block.
   a. control
   b. flex
   c. bed
   d. cold

_____ 32. Tooling of masonry joints improves their appearance and
   a. forces the mortar tight against the block.
   b. requires the use of an electric grinder.
   c. requires the use of a very sharp, pointed chisel.
   d. forces out all visible mortar from the front joints.

_____ 33. A solid masonry course cap is required for foundation walls made of hollow concrete block. One reason for this requirement is that the cap acts as a(n)
   a. thermal barrier.
   b. termite barrier.
   c. finish course.
   d. anchor course.

Carpentry & Building Construction Instructor Resource Guide
Copyright © Glencoe/McGraw-Hill

Concrete Flatwork

MATCHING

Directions: Match each item to the statement or sentence listed.

Column I

_____ 1. Small device left in place as concrete is poured to support the wire fabric at a particular height.

_____ 2. Uniform and shallow layer of material used for fill in the subgrade.

_____ 3. Colorless, odorless, radioactive gas given off by some soils and rocks.

_____ 4. The first step in finishing concrete flatwork.

_____ 5. Finishing step that produces a rounded edge on the slab to prevent chipping or damage.

_____ 6. Helps keep ground moisture from wicking into the slab.

_____ 7. Foundation slab poured between foundation walls.

_____ 8. Foundation preferred where frost is not a problem and where termite infestations are common.

_____ 9. Finely crushed or powdered materials to be removed from fill.

_____ 10. Used when hand-troweling or when floating a large surface to keep the finisher from stepping onto the fresh concrete.

Column II

a. fines
b. chair
c. edging
d. kneeboards
e. subgrade
f. independent slab
g. screeding
h. monolithic slab
i. radon
j. perimeter drain
k. lift

MULTIPLE CHOICE

Directions: In the blank provided, write the letter of the answer that best completes the statement or answers the question.

_____ 11. In places where the ground freezes fairly deep during winter, a(n) _?_ foundation slab is usually poured.
    a. monolithic
    b. independent
    c. unified
    d. thickened-edge

(Continued on next page)

CHAPTER 15 TEST (Continued)

_____ 12. To allow for good drainage, how many inches above the ground should the top of the slab sit?
   a. 8"
   b. 6"
   c. 3½"
   d. 4"

_____ 13. Fill layer consisting of coarse slag, gravel, or crushed stone no more than 2" in diameter is called
   a. subgrade.
   b. a lift.
   c. a fine.
   d. subbase.

_____ 14. The last step in finishing flatwork is
   a. troweling.
   b. floating.
   c. edging.
   d. bullfloating.

_____ 15. To ensure hydration will continue after finishing concrete flatwork,
   a. place concrete in insulated forms.
   b. add more water to the concrete mix.
   c. small smooth pebbles can be scattered in to fresh concrete.
   d. the concrete should be kept moist for at least two days.

_____ 16. A step in the finishing process of flatwork that evens out high or low spots is
   a. edging.
   b. bullfloating.
   c. floating.
   d. screeding.

_____ 17. Which of the following might increase tensile strength and reduce cracking in a foundation slab?
   a. use of bearing wall supports
   b. concrete formwork presets
   c. addition of metal reinforcement
   d. foam insulation in slab base

_____ 18. Calculate the exact number of cubic yards of concrete needed for a basketball court that is 4" thick, 24 ½' wide, 36' long.
   a. 10.878
   b. 130.66
   c. 16.33
   d. 8.166

_____ 19. A disadvantage of a vapor barrier is that it forces moisture in the fresh concrete to escape through the exposed top surface. How can this be avoided?
   a. Pour concrete over a thick layer of sand spread over the vapor barrier.
   b. Build a sub-slab ventilation system made of fill and slag.
   c. Use rigid nonabsorbent extruded or expanded polystyrene.
   d. Chemically heat the sand before placing the slab.

_____ 20. To more precisely position wire fabric reinforcement, support it on
   a. kneeboards.
   b. floats.
   c. chairs.
   d. lifts.

Wood As a Building Material

MATCHING

Directions: Match each lettered item to the numbered description.

Column I

_____ 1. The layer of cells that produces sapwood.

_____ 2. The part of the tree that gives strength and rigidity.

_____ 3. Defect in lumber identified as the presence of bark on the edges.

_____ 4. Deciduous trees such as walnut, mahogany, and maple.

_____ 5. Coniferous trees such as pine, hemlock, and cedar.

_____ 6. Method of sawing lumber from a log lengthwise, parallel to the grain.

_____ 7. Method of sawing lumber from a sectional portion of a log.

_____ 8. Defect in lumber identified as a lengthwise separation.

_____ 9. Produces an average moisture content of 19 percent or less.

_____ 10. Produces lumber with less than 10 percent moisture content.

Column II

a. flat sawing
b. kiln drying
c. cambium
d. quarter sawing
e. air drying
f. heartwood
g. shake
h. wane
i. softwood
j. knothole
k. hardwood

(Continued on next page)

CHAPTER 16 TEST (Continued)

MULTIPLE CHOICE

Directions: In the blank provided, write the letter of the answer that best completes the statement or answers the question.

_____ 11. Lumber grade is an indication of
   a. its quality.
   b. its type.
   c. its weight.
   d. its age.

_____ 12. Which of the following is **not** an advantage of quarter-sawn lumber?
   a. It has a more durable surface.
   b. It holds paints and finishes better.
   c. The boards are of greater width than other cuttings.
   d. It has a low tendency to warp, shrink, or swell.

_____ 13. Wood decay is
   a. caused by certain fungi.
   b. caused by low temperatures.
   c. present in all softwood.
   d. found in dry locations.

_____ 14. Naturally decay-resistant woods include the
   a. sapwood of all walnut, maple, and birch trees.
   b. cambium of yellow and white pine.
   c. heartwood of bald cypress, redwood, and cedars.
   d. summerwood of fir, spruce, and yew.

_____ 15. Which of the following wood-infesting insects do **not** eat the wood?
   a. carpenter ants
   b. dry-wood termites
   c. subterranean termites
   d. powderpost beetles

_____ 16. Which wood-infesting insects are the most destructive?
   a. carpenter ants
   b. subterranean termites
   c. dry-wood termites
   d. deathwatch beetles

_____ 17. You should use preservative-treated wood when building
   a. fireplaces.
   b. cabinets.
   c. decks.
   d. furniture.

_____ 18. What is the moisture content of a block of wood that weighed 75 lbs. when cut and 60 lbs. after drying? **Hint:** Divide the weight of the water in the original piece by the weight of the block after drying.
   a. 80 percent
   b. 25 percent
   c. 28 percent
   d. 20 percent

_____ 19. Softwood with a moisture content (MC) of more than 19 percent is considered _?_ lumber.
   a. dry (seasoned)
   b. wet (select)
   c. green (unseasoned)
   d. natural (dry)

_____ 20. Examining _?_ can identify the early wood and late wood growth of a tree.
   a. the annual rings
   b. the heartwood
   c. the sapwood
   d. the cambium

Engineered Lumber

MATCHING

Directions: Match each item to the statement or sentence listed below.

Column I

_____ 1. Engineered lumber product that pound-for-pound is stronger than a steel beam.

_____ 2. A closely spaced series of wedge shaped cuts made to join lengths of solid wood.

_____ 3. Length of engineered stock that have the same depth as the I-joists of the floor system.

_____ 4. Slight upward curve of a beam measured in inches or radius of curvature.

_____ 5. Can be substituted for solid wood lumber in nearly every instance.

_____ 6. Engineered lumber product made with layers of bonded wood commonly used for the I-joist.

_____ 7. Common types include joist hangers, I-joist hangers, and ties and straps.

_____ 8. The horizontal members on an I-joist.

_____ 9. Used where appearance is very important and all knot holes and voids are filled.

_____ 10. Material resulting when glued sheets of veneer are fed into a machine that uses heat and pressure to cure the adhesive.

Column II

a. web
b. glulam
c. billet
d. engineered lumber
e. finger joint
f. premium grade glulams
g. metal framing connector
h. rim boards
i. flange
j. laminated-veneer lumber
k. architectural grade glulams
l. camber

(Continued on next page)

CHAPTER 17 TEST (Continued)

MULTIPLE CHOICE

Directions: In the blank provided, write the letter of the answer that best completes the statement or answers the question.

_____ 11. Which of the following statements about engineered lumber is true?
  a. Overall performance is highly unpredictable.
  b. It uses wood that might otherwise be wasted.
  c. It is available in limited dimensions.
  d. Its defects are similar to solid wood but is made stronger.

_____ 12. One advantage of using engineered lumber over solid lumber is that it
  a. doesn't shrink and swell as much.
  b. is heavier than solid lumber.
  c. is available in limited lengths.
  d. can be stored outside without worry of moisture damage.

_____ 13. The easiest tool to use when crosscutting an LVL I-joist is a
  a. circular saw.
  b. jig saw.
  c. hand saw.
  d. radial arm saw.

_____ 14. When installing LVL I-joists used in floor construction
  a. never drive nails into the flange at a 45° angle.
  b. always use temporary braces called web stiffeners.
  c. secure them by toenailing through the lower flange.
  d. drive nails sideways into the upper flange.

_____ 15. To encourage air circulation and prevent contact with ground moisture, it is best to store LVL I-joists on
  a. their sides.
  b. stickers.
  c. their ends.
  d. the wrapping.

_____ 16. What engineered lumber product used for columns and studs is made from wood veneer ribbons?
  a. parallel-strand lumber
  b. laminated-veneer lumber
  c. laminated-strand lumber
  d. glue laminated lumber

_____ 17. The most common metal framing connector is the
  a. framing tie.
  b. metal post base.
  c. hurricane clip.
  d. joist hanger.

_____ 18. The most common mistake made when installing joist hangers is to
  a. use too few nails.
  b. use drywall screws.
  c. use too many nails.
  d. install in the wrong direction.

_____ 19. What engineered lumber product is sometimes manufactured with a camber?
  a. LVL I-joist
  b. LVL header
  c. run board
  d. glulam

_____ 20. When I-joists are attached to an intersecting I-joist, the joist hangers should be
  a. bonded.
  b. backed.
  c. laminated.
  d. webbed.

Engineered Panel Products

MATCHING

Directions: Match each item to its definition listed below.

Column I

_____ 1. Made from small flakes of wood bonded together with adhesive.

_____ 2. Very thin sheets of wood sliced from a log.

_____ 3. High-density fiberboard used for interior paneling and cabinet backing.

_____ 4. Dense, non-combustible board often used as lap siding.

_____ 5. Respiratory illness resulting from breathing fine particles of fiber cement.

Column II

a. wood veneer
b. particleboard
c. silicosis
d. fiber-cement board
e. medium-density fiberboard
f. hardboard

Directions: Match each plywood designation with its description, below.

Column I

_____ 6. Face is high-quality veneer; back is lower quality.

_____ 7. Highest quality veneers used on face and back.

_____ 8. Indicates a water-resistant coating was applied.

_____ 9. Has one smooth side and one rough side, could be roof sheathing.

_____ 10. Made with water-resistant glue but isn't weatherproof.

Column II

a. X
b. C-D
c. A-C
d. Exposure-1
e. A-A
f. D-D

MULTIPLE CHOICE

Directions: In the blank provided, write the letter of the answer that best completes the statement or answers the question.

_____ 11. The plywood most commonly used in building construction is made from
   a. softwood.
   b. hardwood veneer.
   c. hardwood.
   d. mixed hard wood and plastics.

(Continued on next page)

CHAPTER 18 TEST (Continued)

_____ 12. Plywood is constructed of
   a. 2 or 4 layers of real wood veneer.
   b. 2, 3, or 4 layers of veneer or strand board.
   c. 3, 5, or 7 layers of real wood veneer.
   d. 2 layers of wood veneer or particleboard.

_____ 13. Composite panel products are constructed of
   a. 2 or 4 layers of wood veneer.
   b. wood pieces mixed with adhesives.
   c. hardwood veneer faces with particleboard cores.
   d. hardwood veneer over metal cores.

_____ 14. How can you determine the grade of a piece of plywood?
   a. Use product information brochures provided by the seller.
   b. Compare the plywood to a standard grading chart.
   c. You need to be able to identify them on sight.
   d. The grade should be stamped on the plywood.

_____ 15. Most plywood panels are 4' wide and
   a. 4' long.
   b. 16' long.
   c. 8' to 12' long.
   d. 4' to 10' long.

_____ 16. A plywood panel intended for foundation use would very likely be
   a. treated with preservative chemicals.
   b. a different color than other plywood.
   c. finished very smoothly on both sides.
   d. heavier in weight than other plywood.

_____ 17. Which of the following statements most accurately describes the characteristics of a 4' × 10' piece of plywood?
   a. The width is 10' and the grain runs lengthwise.
   b. The width is 4' and the grain runs widthwise.
   c. The width is 10' and the grain runs widthwise.
   d. The width is 4' and the grain runs lengthwise.

_____ 18. When cutting plywood with a power saw, properly support the plywood to prevent the
   a. saw from kicking back and causing injury.
   b. plywood from splintering on the good side.
   c. release of fine particle dust into the air.
   d. plywood from warping or cracking.

_____ 19. Because engineered panel products are made with adhesives, it is very important to use __?__ when you sand or machine them.
   a. saw lubricants
   b. very fine sandpaper
   c. a dust mask
   d. protective gloves

_____ 20. Which of the following does **not** accurately describe oriented-strand board (OSB)?
   a. It is generally made with a waterproof adhesive.
   b. It can take long-term exposure to weather elements.
   c. All the strands in each layer run in one direction.
   d. It is made of 3 to 5 layers that are perpendicular to each other.

Framing Methods

MATCHING

Directions: Match each word or phrase to the statement or sentence.

Column I

_____ 1. Used by architects and engineers to predict the performance of any part of a wood frame.

_____ 2. Examples are books, furniture, and appliances.

_____ 3. Examples are paneling, fixtures, and cabinets.

_____ 4. Designed to resist *lateral* movements as in earthquakes.

_____ 5. Used by carpenters to find the right spacing for different sizes of joists or rafters.

_____ 6. Consists of 3 ½" expanded polystyrene foam insulation between sheets of plywood.

_____ 7. The spacing from the centerline of one joist to the centerline of another.

_____ 8. A force that creates stress on a structure.

_____ 9. The spacing between the joists, studs, and rafters is the same.

_____ 10. Fastened to wood framing to give it more strength and stiffness.

Column II

a. load
b. horizontal sheathing
c. design value
d. shear wall
e. sheathing
f. structural insulated panel
g. live load
h. in-line framing
i. dead load
j. span table
k. on center

MULTIPLE CHOICE

Directions: In the blank provided, write the letter of the answer that best completes the statement or answers the question.

_____ 11. What produces tension stresses in the wood farthest from the load and compression stresses closest to the load?
  a. tension
  b. bending
  c. compression
  d. shear

(Continued on next page)

CHAPTER 19 TEST (Continued)

_____ 12. Stresses affecting wood fibers uniformly along the full length of the wood typically in studs, posts, and columns are known as
   a. horizontal shear
   b. compression perpendicular to the grain.
   c. compression parallel to the grain.
   d. tension parallel to the grain.

_____ 13. Which type of stress might occur in a floor joist attached to two walls that are bowing outward?
   a. compression parallel to the grain
   b. horizontal shear
   c. compression horizontal to the grain
   d. tension parallel to the grain

_____ 14. What kind of stresses occurs where two portions of the wood are trying to slide past each other in opposite directions?
   a. horizontal shear
   b. compression perpendicular to the grain
   c. bending
   d. tension parallel to the grain

_____ 15. Which of the following is **not** a dead load?
   a. dry wall
   b. cabinets
   c. furniture
   d. paneling

_____ 16. To use a span table, what three determinations must first be made?
   a. species and grade of wood, spacing on-center, lumber dimensions
   b. live load, species and grade of wood, spacing on-center
   c. the span, live load, spacing on-center
   d. the span, live load, species and grade of wood

_____ 17. Which of the following statements is true about balloon-frame construction?
   a. It is easily adapted to prefabrication.
   b. The studs run from the sill of the foundation to the top plate to the second floor.
   c. Each level of the house is constructed separately.
   d. It does not require long lengths of lumber.

_____ 18. Which of the following statements is true about platform-frame construction?
   a. Each level of the house is constructed simultaneously.
   b. It is the most common method used for one and two story houses.
   c. It is less affected by expansion and contraction.
   d. It has studs that run from the sill attached to the foundation to the top plate of the second floor.

_____ 19. Which of the following is **not** an advantage of structural insulated panels?
   a. The panels can be easily wired.
   b. The panels are quickly erected.
   c. The panels are very efficient.
   d. The panels make the house very strong.

_____ 20. In standard platform-frame construction, wall studs are commonly spaced how far apart?
   a. 12" OC
   b. 19.2" OC
   c. 24" OC
   d. 16" OC

Floor Framing

MATCHING

Directions: Match each item to the statement or sentence listed.

Column I

_____ 1. Used in place of lumber joists where long spans are required.

_____ 2. A large, principal horizontal member used to support floor joists.

_____ 3. A wood or steel vertical member that provides support for a girder.

_____ 4. Round steel posts preferred in residential construction.

_____ 5. A supporting member that projects into space and is supported only at one end.

_____ 6. A floor joist interrupted by a header.

_____ 7. Used to form sides of a large opening.

_____ 8. Method of bracing between long joists.

_____ 9. A length of plywood, OSB, or LVL that ties the ends of the I-joists together.

_____ 10. A block of plywood or OSB sheathing that can be nailed to both sides of an I-joist to improve load-bearing characteristics.

Column II

a. rim board
b. trimmer joist
c. post
d. bridging
e. Lally columns
f. mudsill
g. web stiffeners
h. girder
i. trusses
j. tail joist
k. cantilever

(Continued on next page)

CHAPTER 20 TEST (Continued)

MULTIPLE CHOICE

Directions: In the blank provided, write the letter of the answer that best completes the statement or answers the question.

_____ 11. Which of the following statements is correct about installing posts?
   a. The standard post size is 2×4.
   b. It is best to use composite posts.
   c. Posts should have flat ends.
   d. One post can carry loads for large areas.

_____ 12. What is the first step when installing a sill plate?
   a. Check that the foundation is level and square.
   b. Place the sill stock around the foundation.
   c. Apply the sill sealer to prevent cold air leaks.
   d. Establish the location of sill plate.

_____ 13. When laying out joists, how far from the corner of the sill plate should the first joist be located?
   a. 16"
   b. 24"
   c. 14 ¼"
   d. 15 ¼"

_____ 14. To accommodate extra weight under load-bearing walls, the joists should be
   a. supported by headers.
   b. supported by trimmer joists.
   c. doubled.
   d. set as tail joists.

_____ 15. Headers and trimmer joists are used to create large openings for
   a. stairwells and chimneys.
   b. bearing walls.
   c. bathroom floors.
   d. second-story floors.

_____ 16. Which of the following is **not** a correct statement about subfloor installation?
   a. The grain of the plywood runs at right angles to the joists.
   b. Leave at least a ¾" gap between each panel sheet.
   c. Snap a chalk line to ensure the panels align.
   d. Drive enough nails to hold the panels in place.

(Continued on next page)

Name _____ Date _____ Class _____

CHAPTER 20 TEST *(Continued)*

_____ 17. Chords, webs, and connector plates are three basic parts of a
   a. girder floor system.
   b. panel subfloor.
   c. floor truss.
   d. double layer floor.

_____ 18. When using girder floor framing, what supports the posts?
   a. parallel chord floor trusses
   b. concrete pier footings
   c. box sill plates
   d. metal strap cross bridging

_____ 19. What is a cost effective and efficient way to stiffen a floor and prevent joists from twisting?
   a. Install web stiffeners.
   b. Lay a separate subfloor.
   c. Use floor trusses.
   d. Install cross bridging.

_____ 20. What is the outermost curve of an edgewise bow on a joist?
   a. crown
   b. bridging
   c. web
   d. cantilever

_____ 21. Estimate the number of 4×8 floor sheathing panels that will be required for a two-story house measuring 52' by 30' on each floor.
   a. 97
   b. 98
   c. 49
   d. 47

_____ 22. What are made of wood or steel and are generally placed halfway between, and parallel to, the longest foundation walls?
   a. girders
   b. sill plates
   c. floor joists
   d. subflooring

(Continued on next page)

CHAPTER 20 TEST (Continued)

_____ 23. What provides a smooth bearing surface for floor joists and serves as a connection between the foundation wall and the floor system?
   a. girders
   b. sill plates
   c. subflooring
   d. floor joists

_____ 24. What lends bracing strength to the building, acts as a barrier to cold and dampness, and provides a solid base for the finish floor?
   a. floor joists
   b. girders
   c. sill plates
   d. subflooring

_____ 25. What distributes the load to the foundation walls and provides a nailing surface when sheathing is attached?
   a. subflooring
   b. sill plates
   c. floor joists
   d. girders

Wall Framing and Sheathing

MATCHING

Directions: Match each type of wall framing member to its description below.

Column I

_____ 1. Horizontal lumber that supports a window.

_____ 2. Primary vertical framing member.

_____ 3. Beam across the top of an opening.

_____ 4. Primary horizontal framing member.

_____ 5. Studs installed below a rough sill.

_____ 6. Stud that supports a header over a window or door.

_____ 7. Full-length stud on either side of an opening.

Column II

a. plate
b. header
c. rough sill
d. sheathing
e. cripple stud
f. stud
g. trimmer stud
h. king stud

Directions: Match each type of special wall framing with its description.

Column I

_____ 8. Built to withstand high stresses such as those caused by earthquakes.

_____ 9. Tops of studs are cut on an angle to match the roof's pitch.

_____ 10. May require beveled framing and two headers.

_____ 11. Covers a gap from cabinets up to the ceiling.

_____ 12. Made with blocks to allow direct mounting of inset cabinets.

_____ 13. Made with curved plates cut from plywood.

Column II

a. rake wall
b. angled bay window
c. radius wall
d. rough sill
e. cabinet framing
f. shear wall
g. cabinet soffit

(Continued on next page)

CHAPTER 21 TEST (Continued)

MULTIPLE CHOICE

Directions: In the blank provided, write the letter of the answer that best completes the statement or answers the question.

_____ 14. In laying out wall locations, the first step is to lay out
  a. the positions of all openings, including windows and doors.
  b. the angles for all exterior corner posts.
  c. the location of two intersecting exterior walls.
  d. the stud locations.

_____ 15. If you use the "3-4-5 rule," which of these sets of wall measurements forms a 90° angle?
  a. 6' × 8' with 9.5' between them
  b. 18' × 24' with 29' between them
  c. 15' × 21' with 27' between them
  d. 12' × 16' with 20' between them

_____ 16. Where should an X be placed to mark a plate's location relative to an interior wall?
  a. on the side of the line where the plate will be located
  b. on the very center of the line
  c. on the side of the line opposite where the plate will be located
  d. on the side nearest a bearing wall

_____ 17. When all planned wall lines have been snapped onto a subfloor or slab, the first wall members cut are the exterior
  a. studs.
  b. wall plates.
  c. headers.
  d. butt walls.

_____ 18. To lay out a wall's first opening, measure from __?__ to the center of the opening.
  a. the nearest stud
  b. the centers of the nearest studs on each side
  c. both sides of the opening
  d. one corner of the building

_____ 19. How can you determine the rough opening size required for a door?
  a. Take the width and length measurements directly from the door.
  b. Check standard measurements used for all rough openings.
  c. Check the door schedule in the building plans.
  d. Use a carpenter's layout template to mark the plate.

_____ 20. What should you find at every vertical joint between 4'-wide sheathing panels?
  a. by-wall measurement
  b. header
  c. stud
  d. plate intersection

(Continued on next page)

Name _____ Date _____ Class _____

CHAPTER 21 TEST (Continued)

_____ 21. Information about the location and size of window headers, sills, and door headers is often shown on a
   a. window schedule.
   b. story pole.
   c. door schedule.
   d. floor plan.

_____ 22. You must add plywood spacers to 2× members to make a 3 ½" header
   a. to allow for expansion and contraction in extreme temperatures.
   b. because lumber used for windows and doors is thicker than header lumber.
   c. to bring the level of a window or door up to the height of the trimmer studs.
   d. because 2× members are only 1 ½" thick.

_____ 23. How can you estimate the number of 16" OC studs needed for exterior walls?
   a. Figure one stud per every 16 lineal inches of wall.
   b. Figure one stud per lineal foot of wall.
   c. Figure one stud per every 8 lineal inches of wall.
   d. Figure one stud per every 32 lineal inches of wall.

_____ 24. For walls with a double top plate, how do you figure the total length of top and bottom plates needed?
   a. Multiply the wall length by three.
   b. Multiply the wall length by two.
   c. Divide the wall length by three.
   d. Divide the wall length by two.

_____ 25. How can you determine the number of 4×8 plywood sheets needed to sheathe a house?
   a. Determine the total square footage of floor space and divide by 32.
   b. Add together the lengths of all 4 walls and divide by 12.
   c. Determine the total wall area and divide by 32.
   d. Multiply the house's length by its width and divide by 8.

_____ 26. If you have very little help available to raise the walls while building, it would be better to apply the sheathing
   a. with the face grain going across the studs.
   b. with the face grain going parallel to the studs.
   c. while the framed wall sections are lying down flat.
   d. to the wall studs after they have been erected.

_____ 27. After a wall frame is assembled it should be _?_ by running a tape measure across diagonally opposite corners.
   a. plumbed
   b. braced
   c. squared
   d. sheathed

(Continued on next page)

CHAPTER 21 TEST (Continued)

_____ 28. When you ? a wall, it means you make sure it is perpendicular (at a right angle) to the subfloor.
   a. plumb
   b. square
   c. partition
   d. block

_____ 29. What should be your first step for putting up interior walls in a new building?
   a. Complete any single room and go on to another one.
   b. Erect two short parallel partitions and begin working between them.
   c. Begin work on framing around doorways.
   d. Start by erecting the longest center partition.

_____ 30. Why do builders use a double plate above the top plate of a wall?
   a. to add strength from a better quality of material used
   b. to add strength with this thicker piece of plate
   c. to tie the walls together at the top
   d. to tie the windows together at the top

_____ 31. If a wall supports weight from portions of the house above it, it is called a(n) ? wall.
   a. partition
   b. bearing
   c. divider
   d. interior

_____ 32. Why do some building codes require that sheathing be applied vertically near corners of a building?
   a. to allow for extra insulation and soundproofing
   b. to maintain a consistent outward appearance
   c. to ensure greater rigidity in the structure
   d. to allow for fastening of certain types of siding

_____ 33. Which of these statements is true of power fasteners such as nailing guns?
   a. They pose no safety risks.
   b. They are impossible to use incorrectly.
   c. They apply fasteners more tightly than hammering.
   d. They apply all fasteners with the same amount of force.

Basic Roof Framing

MATCHING

Directions: Match each item shown in Fig. 22-1 to the correct term.

_____ 1. gable roof

_____ 2. hip roof

_____ 3. low-slope roof

_____ 4. shed roof

_____ 5. gambrel roof

_____ 6. mansard roof

_____ 7. Dutch hip roof

Fig. 22-1

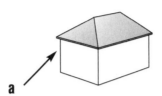
a

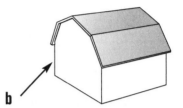
b

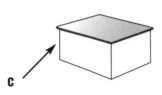
c

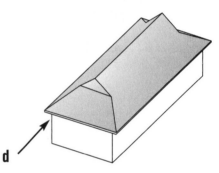
d

e

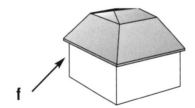
f

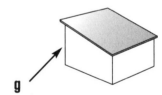
g

(Continued on next page)

CHAPTER 22 TEST (Continued)

Directions: Match each item to the statement or sentence listed below.

Column I

_____ 8. The horizontal piece that connects the upper ends of the rafters.

_____ 9. A notch made in a rafter with an overhang so the rafter will fit against a plate.

_____ 10. Serves the same purpose in a roof as joists do in the floor.

_____ 11. Part of the rafters that extends beyond the exterior walls to form the overhang (eave).

_____ 12. The distance between the outer edges of the top plates.

_____ 13. One-half the span.

_____ 14. Number of inches that a roof rises for every 12" of run.

_____ 15. Vertical distance from the top of the top plate to the upper end of the measuring line.

_____ 16. Need along with unit rise for calculating slope.

Column II

a. total rise
b. total run
c. ceiling joists
d. span
e. unit run
f. bird's mouth
g. unit rise
h. tail
i. rafters
j. ridge board

(Continued on next page)

Name _____ Date _____ Class _____

CHAPTER 22 TEST (Continued)

MULTIPLE CHOICE

Directions: In the blank provided, write the letter of the answer that best completes the statement or answers the question.

_____ 17. The unit-length method of laying out a common rafter uses
   a. the rafter table on a framing square.
   b. the Pythagorean theorem.
   c. a construction calculator.
   d. framing square to step off the lengths.

_____ 18. Which of the following rafters extend diagonally from the corners of the top plates to the ridge board?
   a. common rafters
   b. hip rafters
   c. jack rafters
   d. cripple jack rafters

_____ 19. What is always the unit run of a common rafter?
   a. 6"
   b. 17"
   c. 12"
   d. 12'

_____ 20. If the span is 28' and the total rise is 7' what is the pitch?
   a. 4
   b. ⅓
   c. ½
   d. ¼

_____ 21. If the total run is 12' and the total rise is 4' what is the slope?
   a. $4/12$
   b. $3/12$
   c. $6/12$
   d. $8/12$

(Continued on next page)

Carpentry & Building Construction Instructor Resource Guide
Copyright © Glencoe/McGraw-Hill

CHAPTER 22 TEST (Continued)

_____ **22.** When developing framing plans for a gable roof, the _?_ determines the location of the ridge board.
  a. top plate
  b. centerline
  c. bird's mouth
  d. tail cut

_____ **23.** When using the step method to lay out a common rafter, the number of times the square is moved is determined by the
  a. unit run.
  b. span.
  c. total rise.
  d. total run.

_____ **24.** Ceiling joists are usually placed across the width of the building and parallel to the
  a. outside bearing wall.
  b. rafters.
  c. ridge board.
  d. interior bearing wall.

_____ **25.** To accommodate homes with larger, more open living spaces, what special ceiling framing may be required?
  a. flush girders
  b. stub joists
  c. gang-cut rafters
  d. lookout rafters

Hip, Valley, and Jack Rafters

MATCHING

Directions: Match each word or phrase to the statement or sentence.

Column I

_____ 1. A rafter that extends from the corner of a building diagonally to the ridge board.

_____ 2. A rafter that forms a depression in the roof.

_____ 3. A shortened common rafter.

_____ 4. Where the end of a hip rafter joins the ridge board at an angle.

_____ 5. To bevel the upper edge of the hip rafter.

_____ 6. To bring the top edge of the hip rafter in line with the upper ends of the jacks.

_____ 7. A roof that intersects the main roof.

_____ 8. A gable dormer with side walls.

_____ 9. A level line.

_____ 10. A plumb cut.

Column II

a. backing the hip
b. doghouse dormer
c. heel cut
d. hip rafter
e. side cut
f. seat cut
g. dropping the hip
h. valley rafter
i. common rafter
j. addition
k. jack rafter

(Continued on next page)

CHAPTER 23 TEST (Continued)

MULTIPLE CHOICE

Directions: In the blank provided, write the letter of the answer that best completes the statement or answers the question.

_____ 11. What is the unit run of a common rafter?
   a. 17 inches
   b. 16.97 inches
   c. 12^2 inches
   d. 12 inches

_____ 12. To calculate the length of a hip rafter, multiply the unit run by the number of feet
   a. on a unit rise of a common rafter.
   b. in the total run of a common rafter.
   c. in the total rise of a common rafter.
   d. in the span of a common rafter.

_____ 13. The unit run of a hip rafter is the _?_ of a right triangle with the shorter sides each equal to the unit run of a common rafter.
   a. hypotenuse
   b. area
   c. rise
   d. perimeter

_____ 14. If the hip rafter is framed against the ridge board using a single side cut, what is the shortening allowance?
   a. the thickness of the ridge board
   b. one-fourth the 45° thickness of the common rafter
   c. one-half the 45° thickness of the ridge board
   d. one-half the 45° thickness of a common rafter

_____ 15. Figuring the length of an equal-span valley rafter is the same as figuring
   a. the length of a common rafter.
   b. the length of a hip rafter.
   c. the length of a jack rafter.
   d. the length of an unequal valley rafter.

_____ 16. If the span of the roof addition is shorter than the span of the main roof and the pitch of the addition is the same as the pitch of the main roof, the addition ridge board will be
   a. the same length as the main ridge board.
   b. at a higher level than the main roof ridge board.
   c. at a lower level than the main roof ridge board.
   d. at the same level as the main roof ridge board.

(Continued on next page)

Name _____ Date _____ Class _____

CHAPTER 23 TEST *(Continued)*

_____ **17.** On a hip roof framing plan, there are lines that indicate the hip rafters. What angle do they form with the edge of the building?
  a. 90°
  b. 50°
  c. 35°
  d. 45°

_____ **18.** When laying out a bird's mouth for a hip rafter, set the body of the square at
  a. 6"
  b. 12"
  c. 17"
  d. 9"

_____ **19.** Rather than lay out and mark each jack rafter individually what can be used to save time?
  a. a pattern
  b. a CADD machine
  c. a construction calculator
  d. a framing square

_____ **20.** When constructing a gable dormer without side walls, what is the dormer ridge board fastened to?
  a. the main ridge board
  b. the common ridge board
  c. the header
  d. the hip rafter

_____ **21.** A jack rafter is a shortened _?_ rafter.
  a. common
  b. hip
  c. valley
  d. dormer

_____ **22.** Which type of jack rafter does not contact either a plate or a ridge board?
  a. valley jack rafter
  b. cripple jack rafter
  c. hip jack rafter
  d. common jack rafter

(Continued on next page)

CHAPTER 23 TEST (Continued)

_____ **23.** Jack rafters have the same spacing and unit rise as
   a. hip rafters.
   b. common rafters.
   c. valley rafters.
   d. dormer rafters.

_____ **24.** What is the best way to figure the total lengths of valley jacks and cripple jacks?
   a. Lay out a roof framing plan.
   b. Figure the lengths of a common rafter.
   c. Use a span table.
   d. Use trial and error.

_____ **25.** When two roofs intersect, framing becomes more complex because what is necessary?
   a. common rafters
   b. hip rafters
   c. valley rafters
   d. jack rafters

Roof Assembly and Sheathing

MATCHING

Directions: Match each item to the statement or sentence.

Column I

_____ 1. Structural ridge that the rafters rest upon.

_____ 2. A roof framing member placed at the intersection of two upward-sloping surfaces.

_____ 3. Should connect to bearing walls at no less than a 45° angle.

_____ 4. Horizontal framing member preventing opposing rafter pairs from spreading apart.

_____ 5. The figure upon which the lengths of gable-end studs are based.

_____ 6. Nonstructural ridge that serves as a bearing surface for opposing pairs of rafters.

_____ 7. Measurement used when determining the length of an intersecting ridge.

_____ 8. Thin sheet of material that prevents water from reaching wood framing.

_____ 9. Diverts water around a chimney and prevents ice from building up on the roof.

_____ 10. Top surface of a sheathed roof.

_____ 11. Commonly used under wood shingles and shakes.

_____ 12. Popular where post-and-beam framing techniques are used.

_____ 13. A rafter that extends from the ridge end to the midpoint on the end wall.

_____ 14. Pair of back-to-back shed dormers.

_____ 15. Commonly used before it was replaced by panel sheathing.

Column II

a. end rafter
b. chimney saddle
c. common difference
d. ridge beam
e. closed sheathing
f. plank sheathing
g. flashing
h. ridge
i. saddlebag dormers
j. roof deck
k. ridge board
l. collar tie
m. gable dormers
n. open sheathing
o. shortening allowance
p. braces

(Continued on next page)

CHAPTER 24 TEST (Continued)

MULTIPLE CHOICE

Directions: In the blank provided, write the letter of the answer that best completes the statement or answers the question.

_____ 16. The type of ridge commonly used when framing low-pitched roofs is the
   a. ridge beam.
   b. ridge board.
   c. nonstructural ridge.
   d. common ridge.

_____ 17. To calculate the actual length of a(n) _?_ roof ridge, you must consider the way in which the rafters are framed.
   a. main gable
   b. equa-span addition
   c. dormer
   d. main hip

_____ 18. The actual length of a(n) _?_ roof ridge is the length of the building measured to the outside edge of the wall framing, plus any overhang.
   a. unequal span addition
   b. equal-span addition
   c. main gable
   d. main hip

_____ 19. When laying out the ridge board, it is essential to confirm the
   a. pitch of the rafters.
   b. length of the rafters.
   c. length of the ridge board.
   d. total rise of the rafters.

_____ 20. To lay out the location of rafters on a hip or gable roof, transfer the locations to the ridge board by matching the ridge board
   a. against the marks on the top plates.
   b. to the location of the ceiling joists.
   c. to the location of the studs.
   d. to the location of the jack rafters.

_____ 21. If collar ties will support a ceiling, they should be installed at _?_ rafter pair.
   a. every fourth
   b. every single
   c. every third
   d. every other

_____ 22. Purlins and braces are used on rafters to
   a. install gable-end studs.
   b. install shed dormers.
   c. hold rafter pairs together.
   d. span greater distances.

(Continued on next page)

CHAPTER 24 TEST (Continued)

_____ 23. What is the common difference in the length of gable-end studs placed 24" OC with an 8" unit rise?
   a. 4"
   b. 16"
   c. 8"
   d. 12"

_____ 24. Dormers are framed after a roof opening has been created and
   a. all common rafters are in place.
   b. before common rafters are in place.
   c. sheathing is in place.
   d. tarpaper has been placed on the sheathing.

_____ 25. When reading the performance rating on panel sheathing used for roofs, the number in front of the slash indicates the
   a. number of square feet in the panel.
   b. thickness of the panel.
   c. maximum spacing of supports for the panel.
   d. number of supports used for the panel.

_____ 26. When installing panel roof sheathing, space nails no more than ? OC at supported ends and edges.
   a. 12"
   b. 8"
   c. 18"
   d. 6"

_____ 27. Using ? sheathing promotes ventilation around wood shingles.
   a. open
   b. plank
   c. closed
   d. panel

_____ 28. A gable-end rafter is nailed to the ? first to keep it from slipping out of position as the ridge is being installed.
   a. ridge board
   b. plate
   c. brace
   d. studs

_____ 29. Hip jacks should be nailed to hip rafters and to the ? with 10d nails.
   a. ridge board
   b. common rafters
   c. plate
   d. ridge beam

(Continued on next page)

CHAPTER 24 TEST (Continued)

_____ 30. To prevent panel roof sheathing from buckling after installation,
   a. lay them with the grain parallel to the rafters.
   b. allow ⅛" between panels.
   c. butt panels tightly against each other.
   d. make sure only the edges are exposed to the weather.

_____ 31. Roof sheathing should have __?__ of clearance from the finished masonry on all sides of a chimney opening.
   a. 3"
   b. ¾"
   c. 2"
   d. ⅛"

_____ 32. Which of the following is a method for estimating the numbers of rafters required for a gable roof with rafters spaced 16" OC?
   a. Divide the roof area by 100 and multiply by 1.083.
   b. Take the length of the building in feet and divide by 24" OC.
   c. Take ¾ of the building's length in feet, add one for the end rafter, and double this figure.
   d. Take ¾ of the building's length in feet, add two for the end rafters, and double this figure.

_____ 33. The most common type of ridge is the
   a. ridge board.
   b. ridge beam.
   c. structural ridge.
   d. ridge rafter.

Roof Trusses

MATCHING

Directions: Match each word or phrase to the statement or sentence.

Column I

_____ 1. The most important truss to brace.

_____ 2. The most popular and widely used light-wood truss.

_____ 3. Used by manufacturers of roof trusses because strength properties are critical.

_____ 4. Used for houses with cathedral ceilings.

_____ 5. Used to assemble heavy trusses.

_____ 6. Part of a truss that makes the assembly rigid.

_____ 7. Length of the bottom chord.

_____ 8. Metal part used to connect wood members of a truss.

_____ 9. The most economical truss for short and medium spans.

_____ 10. The top or bottom outer members of a truss.

Column II

a. chord
b. gable-end truss
c. split-ring connectors
d. Fink truss
e. king-post truss
f. web
g. machine stress-rated lumber
h. strut
i. nominal span
j. scissors truss
k. connector plate

MULTIPLE CHOICE

Directions: In the blank provided, write the letter of the answer that best completes the statement or answers the question.

_____ 11. The basic truss part that is pressed into the wood under hydraulic pressure to splice the joint on either side is a
   a. split-ring connector.
   b. connector plate.
   c. chord.
   d. strut.

_____ 12. The top or bottom outer members of the truss are the
   a. webs.
   b. struts.
   c. chords.
   d. connector plates.

(Continued on next page)

CHAPTER 25 TEST (Continued)

_____ 13. Which of the following statements is true about roof trusses?
   a. They are more expensive than traditional roof framing.
   b. They may be built of stock, metal, or timber.
   c. They require interior bearing walls.
   d. They are made on site for faster installation.

_____ 14. The simplest form of truss used for houses is the
   a. king-post truss.
   b. Fink truss.
   c. scissors truss.
   d. W-truss.

_____ 15. Trusses are commonly designed for what spacing?
   a. 16" OC
   b. 20" OC
   c. 48" OC
   d. 24" OC

_____ 16. When installing trusses, temporary bracing is removed
   a. after a second gable-end truss is in place.
   b. as trusses are fastened to the outside walls.
   c. as sheathing is installed.
   d. when permanent bracing is put into place.

_____ 17. Trusses should be lifted and stored
   a. in a vertical, upright position.
   b. in a horizontal, unsupported stack.
   c. in a horizontal, side-by-side position.
   d. individually.

_____ 18. Permanent continuous lateral bracing for trusses consists of
   a. nominal 2" lumber braced to adjacent trusses.
   b. 2×4 stock nailed to the web or lower chord of each truss.
   c. diagonal bracing called sway bracing.
   d. lumber standoffs anchored to stakes driven into the ground.

_____ 19. When fastening trusses to the outside walls, the system that provides superior resistance against wind uplift is
   a. using screws.
   b. toenailing.
   c. using ring-shank nails.
   d. using metal brackets.

_____ 20. When storing trusses on the job site,
   a. store them in a flat position to lesson bending.
   b. cover them with a tarp to prevent rain damage.
   c. cut the steel bands around the bundles to prevent warping.
   d. tip the banded trusses against standing framework.

Steel Framing Basics

MATCHING

Directions: Match each steel framing tool with its definition or attribute listed below.

Column I

_____ 1. Has a depth-sensitive nosepiece.

_____ 2. Uses a blade with carbide-tipped teeth to cut steel.

_____ 3. Connects steel members, such as studs, to tracks.

_____ 4. Used to make bends in long, flat steel.

_____ 5. Cuts with a hot arc, like a welder's torch.

_____ 6. Uses an abrasive blade to rough-cut steel.

_____ 7. Has high-pressure hoses that supply power to a sharp cutting blade.

Column II

a. press brake
b. framing screw gun
c. chop saw
d. plasma cutter
e. circular saw
f. drywall screw gun
g. pneumatic gun
h. portable hydraulic shears

Directions: Match each steel framing screw shown in Fig. 26-1 to its name.

_____ 8. hex washer

_____ 9. pancake head

_____ 10. modified truss or lath

Fig. 26-1

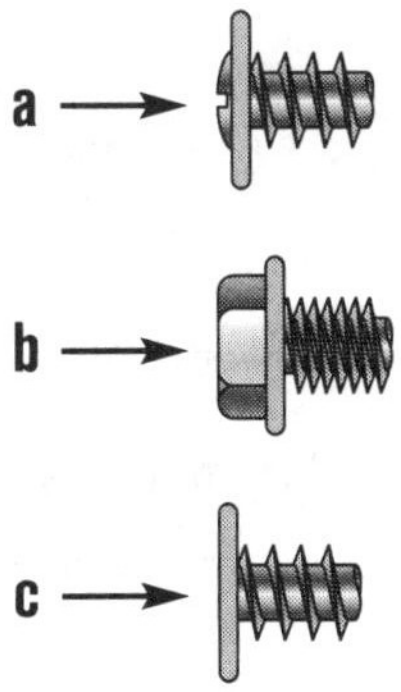

a →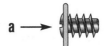

b →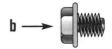

c →

(Continued on next page)

CHAPTER 26 TEST (Continued)

MULTIPLE CHOICE

Directions. In the blank provided, write the letter of the answer that best completes the statement or answers the question.

_____ 11. The type of sheet steel used in residential framing is _?_ steel.
   a. hot-formed
   b. stainless
   c. cold-formed
   d. galvanized hot-formed

_____ 12. Which of the following is **not** true of light-gauge steel framing material?
   a. resists rust and corrosion
   b. is undamaged by insects
   c. contains recycled steel
   d. cannot be recycled

_____ 13. The steel framing design method that uses standardized tables to help a builder select the correct framing members is the _?_ method.
   a. performance
   b. panelized
   c. prescriptive
   d. pre-engineered

_____ 14. The steel framing design method that requires the architect or builder to calculate the size and strength for individual framing members is the _?_ method.
   a. performance
   b. panelized
   c. prescriptive
   d. pre-engineered

_____ 15. In stick-built residential steel framing, studs and joists are spaced on
   a. wider intervals than wood, usually 36" OC.
   b. the standard 16" or 24" OC intervals.
   c. narrower intervals than wood, usually 12" or 14" OC.
   d. intervals divisible by 10, usually 20" or 30" OC.

_____ 16. Panelized construction is used to pre-build
   a. columns, beams, and rafter assemblies.
   b. individual studs, joists, and roof members.
   c. construction templates for pre-engineered components.
   d. flat components such as walls and floors.

(Continued on next page)

CHAPTER 26 TEST (Continued)

_____ 17. Pre-engineered construction is used to pre-build
   a. columns, beams, and rafter assemblies.
   b. individual studs, joists, and roof members.
   c. construction templates for panelized components.
   d. flat components such as walls and floors.

_____ 18. Nailers (or nail guns) are powered
   a. manually.
   b. hydraulically.
   c. pneumatically.
   d. electrically.

_____ 19. Screws are sized according to length and
   a. pullout capacity.
   b. drive type.
   c. number of threads.
   d. diameter.

_____ 20. The most common head style for steel framing screws is the ? head because it provides the most positive drive connection.
   a. modified truss
   b. hex washer
   c. pancake
   d. self-drilling

_____ 21. The gauge of steel generally used to make load-bearing studs for residential construction is
   a. 16 or 17.
   b. 18 or 20.
   c. 22 or 25.
   d. 24 or 26.

_____ 22. In pre-engineered residential steel construction, columns are spaced at
   a. 24" or 36" OC.
   b. 36" or 48" OC.
   c. 3' or 4' OC.
   d. 4' or 8' OC.

(Continued on next page)

CHAPTER 26 TEST (Continued)

_____ **23.** A special hazard that steel building materials can present when left out in the elements is
   a. skin burns from heat or cold.
   b. danger from inhaling oxidized chemicals coming from the steel.
   c. skin burns from galvanizing chemicals if materials get wet.
   d. danger from wet steel reacting with other metals.

_____ **24.** In steel construction, sheathing materials are applied pneumatically to steel framing members with
   a. drive pins and nails.
   b. special steel sheathing screws.
   c. stainless steel rivets.
   d. steel-finishing nails.

_____ **25.** In welding, heat is used to fuse steel pieces; in clinching, __?__ is/are used.
   a. heat, too,
   b. filler metals
   c. pressure
   d. a press brake

Steel Framing Methods

MATCHING

Directions: Match each item to the statement or sentence listed below.

Column I

_____ 1. Load that is carried along the length of a structural member.

_____ 2. Extends beyond the walls on the gabled ends of a house.

_____ 3. Attaches to a structural member to accept a structural load.

_____ 4. C-shape members that attach joists to a foundation.

_____ 5. Aligns vertical and horizontal load-bearing structural members.

Column II

a. clip angle
b. ceiling joist
c. axial load
d. joist load
e. in-line framing
f. joist tracks
g. roof rake

MULTIPLE CHOICE

Directions: In the blank provided, write the letter of the answer that best completes the statement or answers the question.

_____ 6. The first step in laying out steel floor joists is to
  a. make sure each joist is perpendicular to the track.
  b. twist joists to fit between flanges of joist track.
  c. place joist tracks together and mark the locations.
  d. screw through joists and track flanges with #8 screws.

_____ 7. The first step in setting steel ceiling joists (for rafters) is to
  a. install joists at the header one at a time.
  b. mark the layout for the top track.
  c. install blocking every 12 feet OC.
  d. anchor joists at top with #10 screws.

_____ 8. Joists are also called ? because they are attached to a header.
  a. foundation tracks
  b. rim tracks
  c. angle tracks
  d. joist brackets

(Continued on next page)

CHAPTER 27 TEST (Continued)

_____ 9. Steel floor joists should run ? the roof trusses.
   a. perpendicular to the direction of
   b. in alternating vertical planes from
   c. directly below the locations of
   d. in the same direction as

_____ 10. Over each foundation anchor bolt, a ? holds the steel joist track to the foundation.
   a. clip angle
   b. track clip
   c. track anchor
   d. track angle

_____ 11. What type of anchor bolt is put in place before the foundation is poured?
   a. epoxied
   b. concrete
   c. X-head
   d. embedded

_____ 12. What is the result of in-line framing?
   a. The alignment makes it much easier to install the joist tracks.
   b. Axial loads are transferred to the foundation.
   c. Axial loads are distributed evenly among all load-bearing members.
   d. Framing members will be of equal dimensions and placed identically.

_____ 13. When setting steel floor joists in a track, why is it a good idea to leave a ⅛" gap between the track and the end of the joist?
   a. to prevent squeaks
   b. to allow for joist expansion
   c. to allow moisture to run out
   d. to leave room for insulation

_____ 14. The purpose of web stiffeners is to
   a. ensure that corner junctions of joist tracks stay at 90° angles.
   b. reinforce the fastener areas of steel joists.
   c. provide a solid base for sheathing materials on top of joist tracks.
   d. reinforce joists so they don't bend under loads.

_____ 15. Steel wall studs in load-bearing walls must butt tightly inside the track
   a. so they do not slip sideways under typical structural loads.
   b. to prevent loads from being transferred to the foundation.
   c. so that the load is on the stud and not on the screws holding the stud.
   d. to prevent air movement that can cause expansion and contraction.

(Continued on next page)

Name _____ Date _____ Class _____

CHAPTER 27 TEST (Continued)

_____ 16. Which of these is **not** true of framing steel walls?
   a. Full-length framing makes it easier to keep walls straight and square.
   b. Shorter walls require a smaller workforce on site.
   c. With steel, it is relatively easy to keep long walls straight.
   d. To make long walls, short tracks are often spliced together.

_____ 17. When raising steel wall sections, it is important to
   a. attach the C-clamps before raising the wall.
   b. have an adequate on-site workforce to raise the wall.
   c. clamp the sections together with permanent clamps.
   d. force-fit sections tightly into the tracks at the top and bottom.

_____ 18. In nonload-bearing steel frame walls, the top track is attached to
   a. the tops of the wall studs and the end walls.
   b. the wall sheathing and screwed to the studs.
   c. the end walls and an intermediate wall stud.
   d. the ceiling joists or a second-story floor joist.

_____ 19. How should utility holes in steel studs be finished to protect the utility lines?
   a. with short sections of plastic or PVC conduit
   b. with grommets that cover edges and holes
   c. with circular foam "jackets" made for this purpose
   d. with tubular rubber "jackets" made for this purpose

_____ 20. When you're laying out a panelized steel wall, the stud webs should
   a. all face the same direction.
   b. face alternating directions.
   c. match up with the floor joists.
   d. all have C angles attached.

_____ 21. How can you prevent steel wall studs from twisting in the track?
   a. Toenail them in place with small nails made for this purpose.
   b. Install them correctly by butting them in as tightly as possible.
   c. Attach the flanges of each stud securely to both track flanges.
   d. Make sure the tracks are exactly the correct distance apart.

_____ 22. By aligning the punchouts in steel studs, you provide
   a. a straight path for utilities.
   b. space for more insulation.
   c. space for air circulation.
   d. a straight path for reinforcing rod.

(Continued on next page)

CHAPTER 27 TEST (Continued)

_____ 23. Before removing a panelized steel-frame wall from the platform table, you must
   a. apply the sheathing material for strength.
   b. attach the joist tracks to it.
   c. apply insulation along all sides.
   d. check for squareness and then brace.

_____ 24. How are residential steel roof rafters attached to the ridge member?
   a. placed flush with the roof surface and attached with clip angles
   b. placed flush against the ridge member and attached with 2 × 2 inch clip angles
   c. placed perpendicular to the ridge member and attached with angle brackets
   d. placed perpendicular to the ridge member and attached with 2 × 2 inch clip angles

_____ 25. What determines the number and type of connectors needed to fasten steel roof framing to the top plate?
   a. steel gauge of the framing members
   b. the type of roof sheathing to be used
   c. published data on wind loads
   d. the depth and type of insulation

Windows and Skylights

MATCHING

Directions: Match each word or phrase to the statement or sentence.

Column I

_____ 1. The overall size of the window, including casings.

_____ 2. Vertical wood pieces used as separators in combined window units.

_____ 3. A vertical or horizontal piece in a window used to separate panes of glass.

_____ 4. Clear glass or plastic portions of a window.

_____ 5. The part of a window that holds the glass.

_____ 6. An exposed upright member on each side and at the top of the window frame.

_____ 7. Useful in both warm and cool climates to reduce energy flow through the glass.

_____ 8. Useful in cool climates to absorb solar energy.

_____ 9. Inserted into grooves in the outer face of the jamb.

_____ 10. Can be installed without a sash.

_____ 11. Can be made using a combination of two or more materials.

_____ 12. The sash swings outward at the bottom.

_____ 13. The sash swings inward at the top.

Column II

a. sash
b. installation flange
c. hopper window
d. unit dimension
e. jamb
f. awning window
g. glazing
h. low-emissivity glazing
i. mullion strips
j. gas-filled glazing
k. hybrid window
l. muntin
m. heat-absorbing glazing
n. stationary windows

(Continued on next page)

CHAPTER 28 TEST (Continued)

Directions: Match each window shown in Fig. 28-1 to its correct type.

_____ **14.** double-hung

_____ **15.** horizontal sliding

_____ **16.** casement

_____ **17.** awning

_____ **18.** hopper

Fig. 28-1

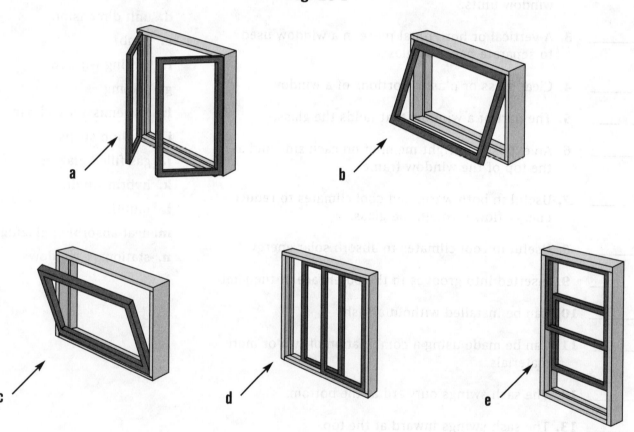

(Continued on next page)

Name _____ Date _____ Class _____

CHAPTER 28 TEST *(Continued)*

MULTIPLE CHOICE

Directions: In the blank provided, write the letter of the answer that best completes the statement or answers the question.

_____ 19. Only half of a _?_ window can be opened at one time.
   a. hopper
   b. stationary
   c. casement
   d. double-hung

_____ 20. The type of window which allows the entire window area to be opened for ventilation is the _?_ window.
   a. hopper
   b. double-hung
   c. casement
   d. awning

_____ 21. To keep air from leaking between the sash and the frame, install _?_ around the window.
   a. cladding
   b. a thermal break
   c. mullion strips
   d. weatherstripping

_____ 22. Which of the following is **not** found on the manufacturer's size table?
   a. size of masonry opening
   b. glass dimensions
   c. combination size
   d. frame size

_____ 23. When installing a window, always check for squareness by
   a. using a plumb.
   b. measuring diagonally across corners.
   c. using a level.
   d. measuring across the top, bottom, and center.

(Continued on next page)

CHAPTER 28 TEST (Continued)

_____ 24. To secure the window in the rough opening, drive a __?__ nail through the flange at one corner.
   a. 1 ½" box
   b. 3" finish
   c. 1 ¾" roofing
   d. 2 ½" common

_____ 25. An egress window must have a(n)
   a. unblocked opening at least 6 sq. ft.
   b. sill height at least 13".
   c. opening height at least 24".
   d. width at least 35".

Residential Doors

MATCHING

Directions: Match each word or phrase to the statement or sentence.

Column I

_____ 1. The most common type of door.

_____ 2. Metal part installed on the door jamb to receive the latch.

_____ 3. Reduces warping and is generally preferred for exterior doors.

_____ 4. Vertical side members of a raised-panel wood door.

_____ 5. Consists of two side jambs and a head jamb.

_____ 6. A light framework of wood or corrugated cardboard faced with thin plywood or hardboard.

_____ 7. Horizontal crosspieces of a raised-panel wood door.

_____ 8. A mortise cut into the edge of a door or jamb.

_____ 9. Opening hardware assembly consisting of knobs, latch, and locking mechanisms.

_____ 10. A panel of glass.

_____ 11. Flexible lengths of materials attached to the edge of the door to close air gaps.

_____ 12. Used when locating and cutting matching gains on the door edge and jamb.

_____ 13. Ensures that the side jambs will be a consistent distance apart.

_____ 14. Fits into a large notch cut into the edge of a door stile.

_____ 15. Door trim.

Column II

a. passage
b. butt-hinge template
c. hollow-core construction
d. bored lock
e. spreader
f. strike plate
g. weatherstripping
h. casing
i. solid-core construction
j. mortise lock
k. rails
l. stiles
m. gain
n. door frame
o. lockset
p. pre-hung
q. light

(Continued on next page)

CHAPTER 29 TEST (Continued)

MULTIPLE CHOICE

Directions: In the blank provided, write the letter of the answer that best completes the statement or answers the question.

_____ 16. Installing a pre-hung door takes approximately ? hours.
   a. 3
   b. 1 ½
   c. 2
   d. 3 ½

_____ 17. The type of hinge most often used for hanging residential doors is the ? hinge.
   a. piano
   b. lap-leaf
   c. flat rabbit
   d. loose-pin butt mortise

_____ 18. The type of lockset most often used on closet doors is the ? lockset.
   a. passage
   b. privacy
   c. entry
   d. mortise

_____ 19. Which of the following statements is true about the proper care and handling of a door?
   a. Seal the wood before installing the door to prevent warping.
   b. Store doors on edge, under cover, and in a clean, dry, well-ventilated area.
   c. Install the door before concrete or plaster has completely dried to condition the door.
   d. Lay doors flat on a dry, level surface to prevent warping.

_____ 20. To determine the hand of a door, stand
   a. facing the hinge jamb and if the door swings left, it is a left-hand door.
   b. with your back against the inside of a closed door. If the door hinges are on the left, it is a left-hand door.
   c. with your back against the hinge jamb and if the door swings to your right, it is a right-hand door.
   d. facing the inside of a closed door and if the knob is on the right, it is a right-hand door.

_____ 21. Which of the following construction features will improve the energy efficiency of exterior doors?
   a. core construction made of rigid insulation
   b. core construction made of corrugated cardboard
   c. flat-panel framework
   d. pre-hung unit construction

(Continued on next page)

Name _____ Date _____ Class _____

CHAPTER 29 TEST *(Continued)*

_____ 22. To improve the energy efficiency when installing an exterior door, use
  a. a bored-hole lockset.
  b. three butt hinges.
  c. a hardwood or metal threshold.
  d. weatherstripping and a storm door.

_____ 23. According to building codes, glazed doors must
  a. have decorative glazing.
  b. be used wherever daylight is needed near a door.
  c. have muntins instead of stiles.
  d. have glazing that is shatter resistant.

_____ 24. Most building codes require that the __?__ be a fire-rated door.
  a. door to the furnace room
  b. door between the house and the attached garage
  c. door to the basement
  d. front door

_____ 25. The type of interior door that exposes only half the opening at one time is the __?__ door.
  a. bifold
  b. folding
  c. sliding
  d. accordion-fold

_____ 26. Exterior doors that come in pairs and allow for unusually wide openings are __?__ doors.
  a. combination storm and screen
  b. sliding glass
  c. French
  d. garage

_____ 27. What clearances are needed between the door jamb and an interior door?
  a. top ½", bottom ½", latch side ⅛", hinge side ⅛"
  b. top ⅛", bottom ⅛", latch side ⅛", hinge side ⅛"
  c. top ⅛", bottom ½", latch side ⅛", hinge side ¹⁄₁₆"
  d. top ¼", bottom ½", latch side ¼", hinge side ½"

_____ 28. Rough openings in the stud walls for interior doors are __?__ than the door.
  a. 2" higher and 2" wider
  b. 3" higher and 2" wider
  c. 2" higher and 3" wider
  d. 3" higher and 3" wider

(Continued on next page)

CHAPTER 29 TEST *(Continued)*

_____ 29. The term "slab door" is another name for a ? door.
   a. flat-panel
   b. pocket
   c. raised-panel
   d. garage

_____ 30. Metal exterior doors have cores that are usually
   a. hollow.
   b. filled with insulating gas.
   c. made of insulation fibers.
   d. made of rigid insulation.

_____ 31. Fire-rated doors are built to resist fire for
   a. 5 minutes.
   b. 10 minutes.
   c. completely.
   d. long enough for occupants to reach safety.

_____ 32. The common minimum width for a bedroom door is
   a. 3'0".
   b. 2'6".
   c. 2'0".
   d. 1'8".

_____ 33. The standard height for stock interior doors is
   a. 7'.
   b. 6'10".
   c. 6'8".
   d. 6'6".

Roof Coverings

MATCHING

Directions: Match each word or phrase to the statement or sentence.

Column I

_____ 1. The amount of weather protection provided by overlapping shingles.

_____ 2. The amount of a shingle showing after installation.

_____ 3. The amount of roofing necessary to cover 100 square feet.

_____ 4. Roofing felt, for example.

_____ 5. Economical roofing product that can be applied quickly over large areas.

_____ 6. Used to cover large flat roofs of commercial buildings.

_____ 7. Causes water to back up beneath the shingles.

_____ 8. Thin metal strip required to protect from water seepage.

_____ 9. Intersection of two roof surfaces where shingles are not applied leaving the flashing exposed.

_____ 10. The amount that adjacent roofing sheets overlap each other horizontally.

_____ 11. Exposed edge of the shingle.

_____ 12. Portion of the shingle not exposed to the weather.

_____ 13. Intersection of two roof surfaces where shingles are interwoven as flashing.

_____ 14. Individual L-shaped pieces of metal used at vertical intersections.

_____ 15. Strips of copper or galvanized steel with a splash diverting rib down the center.

Column II

a. flashing
b. closed valley
c. mineral-surfaced asphalt flashing
d. coverage
e. top lap
f. exposure
g. step flashing
h. square
i. butt edge
j. metal flashing
k. side lap
l. single-ply roofing
m. underlayment
n. open valley
o. ice dam
p. roll roofing

(Continued on next page)

CHAPTER 30 TEST (Continued)

MULTIPLE CHOICE

Directions: In the blank provided, write the letter of the answer that best completes the statement or answers the question.

_____ 16. In recent times, many roofers began to use __?__ instead of shingling hatchets for installing shingles.
   a. roofing hammers
   b. roofing cement
   c. pneumatic nailers
   d. regular hammers and utility knives

_____ 17. Safety guards nailed into rafters that hold boards which prevent workers and tools from sliding off are called
   a. lock-downs.
   b. step flashing.
   c. eave protection.
   d. roof brackets.

_____ 18. Most shingled roofs are sloped
   a. 4-in-12 or greater.
   b. between 3.5-in-12 and 4-in-12.
   c. between 2.5-in-12 and 4-in-12.
   d. below 2.5-in-12.

_____ 19. One type of shingle that is becoming increasingly popular is the laminated or __?__ shingle.
   a. strip
   b. interlocking
   c. architectural
   d. non-interlocking

_____ 20. The first course of shingles should be applied
   a. at the ridge board.
   b. beside the closed valleys.
   c. around the soil stacks.
   d. over the eaves flashing.

_____ 21. The starter strip should be fastened with roofing nails that are placed
   a. 3" or 4" above the eave edge.
   b. 6" above the eave edge.
   c. ⅝" from the soil stacks.
   d. ¼" to ⅜" past the roof edge.

_____ 22. Three-tab shingles require __?__ nails for each strip?
   a. three
   b. four
   c. five
   d. six

(Continued on next page)

CHAPTER 30 TEST (Continued)

_____ 23. Wood shingles and shakes are generally laid over
   a. metal flashing.
   b. roll roofing.
   c. open sheathing.
   d. one layer of underlayment.

_____ 24. What is the difference between wood shingles and shakes?
   a. Shakes are thinner.
   b. Wood shingles are produced largely by hand.
   c. Wood shingles are thinner.
   d. Shakes have a split surface.

_____ 25. Where a chimney or any vertical surface meets a roof, _?_ is used.
   a. step flashing
   b. an adjustable flange
   c. mineral-surfaced asphalt flashing
   d. ribbed metal flashing

_____ 26. Electrolytic corrosion may occur when
   a. copper flashing is used with wood shingles.
   b. one type of metal is used for skylight flashing and another type is used at the eaves.
   c. drip edges are installed over metal eaves.
   d. nails made of stainless steel penetrate valley flashing.

_____ 27. Shingles manufactured in lengths of 24", 18", and 16" and available in three grades are _?_ shingles.
   a. interlocking asphalt
   b. mineral-surfaced asphalt
   c. fiberglass
   d. wood

_____ 28. A common ventilation method that does not rely on fans is _?_ venting.
   a. louvered shingle
   b. passive
   c. hip and ridge
   d. exposure

_____ 29. The process of carrying bundles of shingles to the roof and distributing them is called
   a. loading the bundles.
   b. the starter course.
   c. stacking the roof.
   d. breaking the joints.

(Continued on next page)

CHAPTER 30 TEST (Continued)

_____ 30. In areas where ice dams form along the eaves, solid sheathing should be applied above the eave line and covered with
   a. a double layer of No. 15 asphalt saturated felt.
   b. two layers of tar.
   c. specially treated asphalt shingles.
   d. laminated waterproof paper.

_____ 31. Shakes and wood shingles are graded according to the
   a. type of wood.
   b. thickness of the wood.
   c. quality of the wood.
   d. shape of the shake or shingle.

_____ 32. Shingles designed to resist strong winds are ? shingles.
   a. three-tab
   b. interlocking
   c. laminated
   d. architectural

_____ 33. The single most important step in applying roof shingles is
   a. placement of the starter course.
   b. aligning the courses.
   c. proper nailing.
   d. the installation method.

Roof-Edge Details

MATCHING

Directions: Match each word or phrase to the statement or sentence.

Column I

_____ 1. Consists of a fascia, a soffit, and various types of molding.

_____ 2. Part that provides a transition between the rake and a cornice.

_____ 3. Portions of a roof that project beyond the walls.

_____ 4. A board nailed to the end of the rafters that serves as a mounting surface for gutters.

_____ 5. Extends from the ridge board to a structural fascia to help support the rake section.

_____ 6. Members that form a horizontal surface to which the soffit is attached.

_____ 7. The part of a gable roof that extends beyond the end walls.

_____ 8. The underside of the eaves which can be enclosed or left open.

_____ 9. Short metal tubes placed between the inner and outer faces of the gutter.

_____ 10. Short piece of 2× framing lumber nailed between the roof rafters to seal off the attic space.

Column II

a. ledger
b. cornice return
c. fly rafter
d. frieze block
e. cornice
f. lookouts
g. eaves
h. soffit
i. fascia
j. rake
k. ferrules

MULTIPLE CHOICE

Directions: In the blank provided, write the letter of the answer that best completes the statement or answers the question.

_____ 11. An open cornice consists of
 a. roof sheathing, fascia, and a soffit.
 b. frieze blocks, molding, and a fascia.
 c. a frieze board and one or more pieces of molding.
 d. frieze blocks, a soffit, and roof sheathing.

(Continued on next page)

CHAPTER 31 TEST (Continued)

_____ 12. Which of the following statements is true about a closed cornice?
   a. A higher grade of plywood sheathing needs to be used.
   b. The soffit is nailed to the underside of the rafters.
   c. This type of cornice entirely encloses the rafter tails.
   d. This type of cornice is used on houses with no rafter overhang.

_____ 13. A box cornice is most often constructed
   a. by nailing the soffit to the lookouts.
   b. by nailing frieze blocks between the roof rafters.
   c. by using tongue-and-groove boards as roof sheathing.
   d. without using lookouts.

_____ 14. Before building a box cornice,
   a. determine the length of the lookouts.
   b. install the fascia and gutters.
   c. check the plumb cuts on the rafter tails to make sure they are in line with one another.
   d. temporarily nail a ledger against the wall and the rafters.

_____ 15. On a box cornice, ? is one of the most popular materials used for the soffit.
   a. plywood   c. vinyl
   b. aluminum   d. hardboard

_____ 16. On an open cornice, a curved piece of wood called a pork chop can be attached to the underside of the rake trim to serve as
   a. a frieze block.   c. a rake extension.
   b. flashing.   d. a cornice return.

_____ 17. A gutter system consists of gutters, downspouts, elbows, and
   a. frieze boards.   c. fascia.
   b. leaders.   d. splash blocks.

_____ 18. Metal gutters should slope toward the downspouts at least
   a. 1" every 8'.   c. 1" every 12'.
   b. 1" every 16'.   d. 1" every 14'.

_____ 19. The two general types of gutters are
   a. seamless and continuous.   c. formed-metal and half-round.
   b. round and rectangular.   d. corrugated and smooth.

_____ 20. Wide rake extensions require
   a. a pork chop.   c. the spike and ferrule method.
   b. ladder framing.   d. a frieze board.

Siding

MATCHING

Directions: Match each word or phrase to the statement or sentence.

Column I

_____ 1. Solid wood siding that may use battens to cover joints.

_____ 2. The most common type of solid wood siding.

_____ 3. Sheathing protection made of asphaltic felt.

_____ 4. Sheathing protection made from high-density polyethylene fibers.

_____ 5. Clapboard siding.

_____ 6. Done on site before installation of wood siding to protect and improve durability.

_____ 7. Determined by deducting the minimum overlap from the total width of the siding.

_____ 8. Measuring device made on site to ensure a uniform layout all around the house.

_____ 9. Used where two lengths of siding meet for extra weather tightness.

_____ 10. Used to accurately mark siding that must fit against a vertical surface.

_____ 11. Low-grade layer of wood shingles that will not be exposed to the weather.

_____ 12. Nailed to sheathing when necessary to hold the ends of vinyl siding panels around windows.

_____ 13. Durable cementlike product applied as a finish.

_____ 14. New type of siding material that will not rot, burn, or split and resists mold, mildew, and termites.

_____ 15. The most expensive nails.

_____ 16. Difficult nails to install by hand but can be readily installed by pneumatic nailers.

Column II

a. wood block gauge
b. stucco
c. vertical siding
d. butt joint
e. story pole
f. horizontal siding
g. J-channel
h. aluminum nails
i. back-priming
j. building paper
k. undercourse
l. stainless steel nails
m. scarf joint
n. housewrap
o. fiber cement
p. exposure
q. plain-bevel siding

(Continued on next page)

CHAPTER 32 TEST (Continued)

MULTIPLE CHOICE

Directions: In the blank provided, write the letter of the answer that best completes the statement or answers the question.

_____ 17. To prevent moisture problems, _?_ is used over door and window frames.
   a. undercourse
   b. proper detailing
   c. flashing
   d. high-quality caulking

_____ 18. Never use _?_ between the horizontal courses of beveled siding because moisture vapor must be allowed to escape from behind the siding.
   a. undercourse
   b. proper detailing
   c. flashing
   d. caulking

_____ 19. When maximum weather resistance is required, _?_ nails are recommended.
   a. plated
   b. hot-dipped
   c. stainless steel
   d. aluminum

_____ 20. When installing plain-bevel siding, what should be done first?
   a. Make a story pole.
   b. Nail a furring strip to provide support for the first course.
   c. Snap a chalk line around the perimeter of the house.
   d. Attach the first board to the bottom plate.

_____ 21. The method for installing siding shingles that uses high-grade shingles over undercourse-grade shingles is _?_ coursing.
   a. decorative
   b. ribbon
   c. double
   d. single

_____ 22. A rustic effect can be created by using the _?_ coursing style for installing siding shingles.
   a. double
   b. single
   c. ribbon
   d. decorative

(Continued on next page)

Name _____ Date _____ Class _____

CHAPTER 32 TEST *(Continued)*

_____ 23. When nailing vinyl siding, you should
   a. always drive nails tight.
   b. never drive nails straight but on the angle with the siding.
   c. space fasteners no more than 24" apart.
   d. leave approximately 1/32" between the underside of the nail head and the vinyl.

_____ 24. The type of siding that does not require sheathing is _?_ siding.
   a. fiber cement
   b. plywood
   c. solid wood
   d. vinyl

_____ 25. A NIOSH-approved respirator is recommended when cutting, drilling, or sanding _?_ siding.
   a. stucco
   b. shingle
   c. fiber cement
   d. plywood

_____ 26. Stucco is typically applied over
   a. sheathing.
   b. housewraps.
   c. building paper.
   d. metal lath.

_____ 27. The two kinds of _?_ nails are plated and hot-dipped.
   a. galvanized
   b. stainless steel
   c. high-tensile-strength aluminum
   d. ring-shank

_____ 28. Tight-fitting butt joints are obtained by
   a. cutting a closure board about 1/16" longer than the measured length.
   b. using alternative corner treatment.
   c. staggering the joints.
   d. using 6d siding nails.

_____ 29. Which grade of wood shingles is used for an undercourse?
   a. first grade
   b. second grade
   c. third grade
   d. fourth grade

(Continued on next page)

CHAPTER 32 TEST (Continued)

_____ 30. To secure vinyl siding panels to sheathing, use
   a. a slotted nailing flange at the top.
   b. battens.
   c. zinc-coated or galvanized metal fabric.
   d. tongue-and-groove or shiplap joints.

_____ 31. Nails used for vinyl siding should be
   a. ring-shank nails.
   b. 2" long with heads 3/16" in diameter and a 1/4" shank.
   c. 6d hot-dipped galvanized steel nails.
   d. 1 1/2" long with heads 5/16" in diameter and a 1/8" shank.

_____ 32. The method for installing siding shingles that uses the same grade of shingles for the undercourse as for the outer course is __?__ coursing.
   a. single
   b. double
   c. ribbon
   d. decorative

_____ 33. An advantage to using plywood sheet siding is that it
   a. covers large areas quickly, taking less time to install.
   b. will not burn, rot, or split.
   c. can be easily removed and replaced.
   d. requires little maintenance.

Brick-Veneer Siding

MATCHING

Directions: Match each word or phrase to the statement or sentence.

Column I

_____ 1. A metal bar with a shaped end used to pack the mortar into the joints.

_____ 2. Chisel-like tool used to cut a brick cleanly.

_____ 3. The brick mason's basic tool.

_____ 4. Adding water to a batch of mortar that has become too stiff to work.

_____ 5. Can be standard or oversized and white or yellow.

_____ 6. Has holes in the blade to make mixing easier.

_____ 7. Provides drainage near the bottom of the walls.

_____ 8. Constructed first at both ends of a wall and then the remaining brick is laid between them.

_____ 9. Has a chisel blade and a square face.

_____ 10. Small L-shaped device used to hold a mason's string to align courses.

Column II

a. mortar hoe
b. jointer
c. line block
d. mason's rule
e. brick hammer
f. brick set
g. brick tongs
h. lead corners
i. brick trowel
j. weep hole
k. retempering

MULTIPLE CHOICE

Directions: In the blank provided, write the letter of the answer that best completes the statement or answers the question.

_____ 11. The easiest, quickest, and most accurate way to cut a brick is to use a
   a. masonry hammer.
   b. brick trowel.
   c. masonry saw.
   d. brick hammer and a brick set.

(Continued on next page)

CHAPTER 33 TEST (Continued)

_____ 12. The type of brick that is used primarily for exterior surfaces such as veneer walls is _?_ brick.
   a. fire
   b. facing
   c. refractory
   d. building

_____ 13. Bricks with cores
   a. are more expensive to produce.
   b. are less expensive to ship.
   c. do not harden evenly.
   d. make a weaker mechanical connection than solid bricks.

_____ 14. The size of a modular brick is
   a. a nominal size that includes an allowance for the mortar joint.
   b. its actual dimensions of thickness (height by length).
   c. based on multiples of 5".
   d. called Norman size.

_____ 15. The type of mortar that may be required for areas high in earthquake activity is
   a. Type N.
   b. Type O.
   c. Type M.
   d. Type S.

_____ 16. The type of mortar used as a general-purpose mortar for brick-veneer walls is
   a. Type N.
   b. Type M.
   c. Type S.
   d. Type O.

_____ 17. Small amounts of mortar may be mixed in a wheelbarrow or a
   a. mortar mixer.
   b. power mixer.
   c. mortar box.
   d. mechanical mixer.

_____ 18. Which of the following is true about working in cold weather?
   a. Bricks should be warmed up to 60° before use.
   b. Cold weather slows the hydration process in mortar.
   c. Cold temperatures speed the curing process.
   d. Wet bricks should be discarded if they freeze.

_____ 19. A brick veneer wall must be supported by
   a. corner poles.
   b. refractory bricks.
   c. a masonry or concrete foundation.
   d. sheathing.

_____ 20. To calculate a rough estimate of bricks needed for a brick-veneer wall,
   a. divide the square footage by 7.
   b. multiply the square footage by 7.
   c. divide the square footage by 5.
   d. multiply the square footage by 5.

Stairways

MATCHING

Directions: Match each item to the statement or sentence listed below.

Column I

_____ 1. Part of a stairway on which people step.

_____ 2. The clearance directly above a step.

_____ 3. Another name for a stringer.

_____ 4. Wedge-shaped radiating treads used to turn a stair.

_____ 5. The vertical boards between the treads.

_____ 6. The floor area where a flight of stairs ends or begins.

_____ 7. Finished board nailed to the wall before the stringers are installed.

_____ 8. A stairway that leads to the basement.

Column II

a. winders
b. headroom
c. skirtboard
d. horse
e. tread
f. risers
g. landing
h. service
i. framing
j. joint

Directions: Match each item to the statement or sentence listed below.

Column I

_____ 9. A stairway that has no riser.

_____ 10. Slender vertical members that support the handrail.

_____ 11. Supports the ends of the handrail.

_____ 12. The minimum headroom required by code.

_____ 13. The tread and riser.

_____ 14. The ideal height of a riser.

_____ 15. Parts of a stairway that support the treads.

_____ 16. The proper height for a handrail.

Column II

a. balusters
b. 7 inches
c. step
d. open-riser
e. stringers
f. 6 feet 8 inches
g. newel
h. 34-38 inches
i. stairwell
j. beams

(Continued on next page)

CHAPTER 34 TEST (Continued)

MULTIPLE CHOICE

Directions: In the blank provided, write the letter of the answer that best completes the statement or answers the question.

_____ 17. Most stairways are built inside of a vertical shaft called a
   a. service stairway.
   b. finish stairway.
   c. stairwell.
   d. landing.

_____ 18. What is the most common type of stairway found in houses?
   a. cleat-stringer
   b. open stringer
   c. closed stringer
   d. cut-stringer

_____ 19. What is the vertical distance from the finished surface of one floor to the finished surface of the next floor?
   a. total rise
   b. total run
   c. unit rise
   d. unit run

_____ 20. What is the vertical distance from the top of one tread to the top of the next highest tread?
   a. total rise
   b. total run
   c. unit rise
   d. unit run

_____ 21. When treads are less than 1 ⅛ inches thick or if the stairs are more than 2- 6 feet wide, what needs to be installed?
   a. a third stringer to support the stairway
   b. a kick plate to anchor the stairway
   c. a landing to conserve space
   d. a nosing to extend the tread

_____ 22. The distance from the face of one riser to the face of the next riser is
   a. total rise.
   b. total run.
   c. unit rise.
   d. unit run.

(Continued on next page)

Name _____ Date _____ Class _____

CHAPTER 34 TEST (Continued)

_____ 23. Because of stairway accidents, _?_ codes tightly regulate the design of stairways
   a. handrail
   b. stairway
   c. architectural
   d. building

_____ 24. The purpose of the skirtboard is to
   a. allow the handrail to be attached to the wall with adjustable metal brackets.
   b. protect the wall from damage and make it easier to paint or wallpaper.
   c. be the primary support of a handrail.
   d. support the floor joists that were cut to create the stairwell opening.

_____ 25. Which of the following is the first step in stairway construction?
   a. Calculate the total rise and total run.
   b. Lay out stringers and install treads.
   c. Calculate the unit rise and unit run.
   d. Calculate and lay out the total rise and unit run.

_____ 26. What supports the treads on cleat-stringer stairways?
   a. cleats attached to stringers
   b. notches cut into stringers
   c. joists below the subflooring
   d. interlocking joinery

_____ 27. Calculate the unit rise for a stairway if the total rise is 9 feet 3 inches and there are 16 risers.
   a. 7 ¼ inches
   b. 7 ⅜ inches
   c. 7 ½ inches
   d. 7 ⅛ inches

_____ 28. Calculate the total run for a stairway if there are 18 treads and the unit run is 11 ⅜ inches.
   a. 207 ½ inches
   b. 7 feet 8 inches
   c. 204 ¾ inches
   d. 16 feet 10 inches

_____ 29. What type of stair or stairway is built for occasional attic access?
   a. cut-stringer stairways
   b. hinged stairs
   c. spiral stairs
   d. clear-stringer stairways

(Continued on next page)

CHAPTER 34 TEST (Continued)

_____ **30.** What type of stringers have recesses for wedges mortised into them?
   a. horse stringers
   b. carriage stringers
   c. open stringers
   d. closed stringers

_____ **31.** To allow water to drain and snow to be removed, exterior stairs often do not have
   a. risers.
   b. cut-stringers.
   c. cleat-stringers.
   d. treads that slope.

_____ **32.** Construction time for a stairway is about _?_ hours. This includes rough-cutting the stringers and framing and installing the stringers, treads, and risers.
   a. 3
   b. 9
   c. 15
   d. 25

_____ **33.** As a general rule, what should the sum of one riser and one tread add up to be?
   a. between 10 inches and 12 inches
   b. less than 6 feet 8 inches
   c. between 17 inches and 18 inches
   d. between 34 inches and 38 inches

Molding and Trim

MATCHING

Directions: Match each item to the statement or sentence.

Column I

_____ 1. Usually, a straight length of wood such as 1×4 S4S.

_____ 2. Usually, a narrow length of wood with a shaped profile.

_____ 3. Small offset that allows the carpenter to adjust the fit of a casing.

_____ 4. A marking process that allows a piece of wood to be precisely fit.

_____ 5. The angle at which molding projects away from a wall.

_____ 6. Used when two lengths of molding or trim intersect at an inside corner.

_____ 7. Used when two lengths of molding or trim intersect at outside corners.

_____ 8. Basic molding around a door or window.

_____ 9. Fairly large molding that requires special cutting and installation techniques.

_____ 10. Used where square-edged baseboards meet at inside corners.

Column II

a. scribing
b. casing
c. trim
d. butt joint
e. crown molding
f. compound miter angle
g. finger-jointed molding
h. springing angle
i. reveal
j. coped joint
k. molding
l. miter joint

(Continued on next page)

CHAPTER 35 TEST (Continued)

Directions: Match each item shown in Figs. 35-1 and 35-2 to the correct term.

_____ 11. base shoe

_____ 12. stool

_____ 13. coped joint

_____ 14. apron

_____ 15. miter joint

_____ 16. base cap

Fig. 35-1

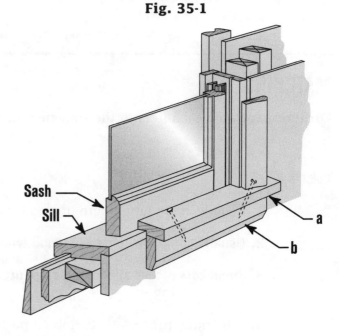

Fig. 35-2

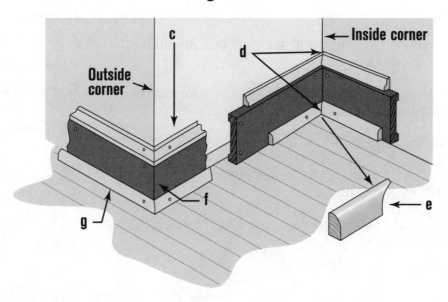

(Continued on next page)

CHAPTER 35 TEST (Continued)

MULTIPLE CHOICE

Directions: In the blank provided, write the letter of the answer that best completes the statement or answers the question.

_____ 17. A practical purpose for baseboard is that it
   a. conceals the large gap between the jams and surrounding framing.
   b. protects the lower portion of a wall during cleaning.
   c. reinforces window jams.
   d. helps control moisture on outside walls.

_____ 18. When scribing a baseboard butt joint to make a tight fit, draw the line and make the cut(s) on
   a. the first installed piece.
   b. the second installed piece.
   c. both pieces before they are installed.
   d. a piece of plywood scrap used to fill the gap.

_____ 19. When coping a baseboard joint, the first step is to
   a. square-cut the end.
   b. use a coping saw and cut a 45° angle on the square-cut end.
   c. miter the end at 45°.
   d. back-cut the mitered end.

_____ 20. Square-edged baseboards should be installed with a _?_ joint at the inside corners.
   a. mitered lap
   b. mitered
   c. coped
   d. butt

_____ 21. When more than one length of molding is needed along a wall, the pieces are joined using a _?_ joint.
   a. mitered lap
   b. butt
   c. coped
   d. mitered

_____ 22. Some manufacturers make trim from short lengths of wood that are _?_ together.
   a. mitered
   b. glued
   c. coped
   d. finger-jointed

(Continued on next page)

CHAPTER 35 TEST *(Continued)*

_____ 23. What kind of molding might be specified for a living room and dining room?
   a. shelf edge
   b. quarter round
   c. cove
   d. crown

_____ 24. Where door casing meets the floor, it is given a ? cut.
   a. mitered
   b. mitered lap
   c. square
   d. coped

_____ 25. Casing with a molded shape must have
   a. mitered corner joints.
   b. compound-miter cuts.
   c. coped inside joints.
   d. square-cut corner joints.

Cabinets and Countertops

MATCHING

Directions: Match each word or phrase to the statement or sentence.

Column I

_____ 1. Also called a galley design, it is often used where space is limited.

_____ 2. Traditional style cabinet; has frame around opening for mounting hardware.

_____ 3. Kitchen storage and workspace separated from other cabinets.

_____ 4. Lower cabinet that supports a countertop.

_____ 5. A layout that is usable and safe for a wide variety of people.

_____ 6. Layout with the sink on a short wall and range and refrigerator against opposite longer walls.

_____ 7. Lower cabinet specifically used in a bathroom.

_____ 8. Arrangement of the three primary work centers in a kitchen.

_____ 9. Upper cabinet in a kitchen or bathroom.

_____ 10. European style cabinet; has concealed hinges and no framing around the opening.

Column II

a. universal design
b. base cabinet
c. work triangle
d. L-shaped kitchen
e. island
f. face-frame cabinet
g. vanity cabinet
h. parallel-wall kitchen
i. wall cabinet
j. frameless cabinet
k. U-shaped kitchen

MULTIPLE CHOICE

Directions: In the blank provided, write the letter of the answer that best completes the statement or answers the question.

_____ 11. The three sides of a kitchen work triangle should add up to
 a. between 9' and 12'.
 b. 10' or less.
 c. between 12' and 26'.
 d. about 30'.

(Continued on next page)

CHAPTER 36 TEST (Continued)

_____ 12. Cabinets that can be built to any size, style, or finish are called _?_ cabinets.
   a. semi-custom
   b. stock
   c. warehouse
   d. custom

_____ 13. The strongest drawer guides are _?_ guides.
   a. side-mounted
   b. center-mounted
   c. top-mounted
   d. hinge-style

_____ 14. On face-frame cabinets, a common type of hinge used is the _?_ hinge which consists of two plates connected with a pin.
   a. cup
   b. barrel
   c. European
   d. knife

_____ 15. To determine the starting point for the layout of base cabinets,
   a. set a corner cabinet in place.
   b. locate and mark the centerlines of the studs.
   c. locate the highest point of the floor in the cabinet area.
   d. locate the lowest point of the floor in the cabinet area.

_____ 16. Typically, the layout line for installing base cabinets is _?_ point of the floor.
   a. 34 ½" above the highest
   b. 36" above the highest
   c. 34 ½" above the lowest
   d. 36" above the lowest

_____ 17. Base cabinets should be mounted with screws that go through the _?_ and into wall studs.
   a. cabinet backing
   b. face frame
   c. back frame
   d. mounting rail

_____ 18. As you install each base cabinet, connect it to the previous one using
   a. a combination of glue and nails.
   b. trim-head screws or bolts.
   c. permanent clamps.
   d. 2 ½" or 3" wood screws.

_____ 19. If cabinets do not quite fill the wall section in which they are installed,
   a. take out the cabinets and reinstall them, evening out the gaps.
   b. return one cabinet and have a new one custom-built to fit.
   c. fill the gap with material that matches the walls.
   d. insert filler strip(s) that match the cabinets.

(Continued on next page)

Name _____ Date _____ Class _____

CHAPTER 36 TEST *(Continued)*

_____ **20.** The separate continuous base often used for _?_ cabinets helps avoid damage to the cabinets when flooring is installed.
   a. factory-built
   b. stock
   c. custom-built
   d. portable

_____ **21.** A wall cabinet should be mounted securely to the wall with a minimum of
   a. eight drywall screws.
   b. four nails driven into the studs.
   c. four round-head screws that extend into the studs.
   d. eight hollow-wall anchors.

_____ **22.** The most common distance between a countertop and the bottom of a wall cabinet is
   a. 15".
   b. 18".
   c. 20".
   d. 22".

_____ **23.** The clearance required between a range top and a wall cabinet directly above it varies according to local building codes, but
   a. is at least 16".
   b. 18" is usually acceptable.
   c. is more than 18".
   d. is usually between 16" and 18".

_____ **24.** If the wall behind a cabinet is very irregular,
   a. scribe and trim the back of the cabinet to fit the irregularity.
   b. cover the gaps with filler strips.
   c. build up or sand the wall until it's even.
   d. use toggle bolts as well as screws.

_____ **25.** During installation, support wall cabinets by
   a. having a coworker hold them.
   b. placing one cabinet underneath another.
   c. using 2 × 4 braces.
   d. using jacks or a plywood stand.

_____ **26.** In a catalog, the product number for a wall cabinet 27" wide and 30" high would be
   a. W3027.
   b. WC2730.
   c. W2730.
   d. WC3027.

(Continued on next page)

CHAPTER 36 TEST (Continued)

_____ 27. Stone, stainless steel, solid surfacing, and ? are installed by contractors specialized in these materials.
   a. ceramic tile
   b. plastic laminates
   c. Corian
   d. Formica

_____ 28. To create a countertop, plastic laminate is usually adhered to a ? with contact cement.
   a. piece of smooth, flat hardwood
   b. metal countertop form
   c. plastic countertop form
   d. substrate base

_____ 29. The grade of laminate used for countertops is the ? laminate.
   a. preformed
   b. vertical surface
   c. postformed
   d. general purpose

_____ 30. The grade of laminate generally used on prefabricated countertop sections is the ? laminate.
   a. preformed
   b. vertical surface
   c. postformed
   d. general purpose

_____ 31. Because laminate dulls tools, use ? cutting tools whenever possible.
   a. power
   b. carbide-tipped
   c. disposable
   d. hand-operated

_____ 32. What procedure should be followed when applying adhesive-coated laminate to a substrate?
   a. Start at the "wall" edge first, pressing across the width.
   b. Begin at the edge that will be the most visible and work across either width or length.
   c. Apply trim pieces first, and then apply sheets starting at the middle of each.
   d. Start from a short end, pressing lengthwise. Apply trim pieces last.

_____ 33. Finish the edges of a plastic-laminated countertop by using a ? or plastic laminate trimmer to remove the excess laminate.
   a. sander
   b. circular saw
   c. router
   d. small hand saw

Wall Paneling

MATCHING

Directions: Match each word or phrase to the statement or sentence.

Column I

_____ 1. The most common type of wall paneling.

_____ 2. The most expensive type of wall paneling.

_____ 3. Used to separate stacked wood products and allow air circulation.

_____ 4. Sometimes referred to as panelboard.

_____ 5. Allowing paneling to become accustomed to the temperature and humidity of the room.

_____ 6. Thick adhesive used to install paneling.

_____ 7. Metal or plastic fitting that is screwed to the front of an electrical outlet box.

_____ 8. Permanently installed to keep paneling from touching another surface.

_____ 9. Technique used to conceal nails when installing vertical paneling.

_____ 10. Paneling that runs partway up the wall from the floor.

Column II

a. stickers
b. furring strip
c. ringshank nail
d. sheet paneling
e. conditioning
f. blocking
g. box extender
h. blind nailing
i. raised paneling
j. medium-density fiberboard paneling
k. wainscoting
l. mastic

MULTIPLE CHOICE

Directions: In the blank provided, write the letter of the answer that best completes the statement or answers the question.

_____ 11. The type of paneling that is made of solid wood with interlocking edges is _?_ paneling.
  a. raised
  b. board
  c. sheet
  d. MDF

(Continued on next page)

CHAPTER 37 TEST (Continued)

_____ 12. Paneling should be stored
   a. indoors, stacked on the floor with stickers to separate the sheets.
   b. outdoors, stacked with stickers to raise sheets off the ground.
   c. indoors, resting on the short edge with stickers to raise sheets off the floor.
   d. outdoors, resting on the long edge with stickers between the sheets.

_____ 13. Paneling should be conditioned for
   a. 48 hours for both sheet and wood paneling.
   b. 7 days for sheet paneling and 48 hours for wood paneling.
   c. 48 hours for wood paneling and 7 hours for sheet paneling.
   d. 48 hours for sheet paneling and 7 days for wood paneling.

_____ 14. The *first* step in wall preparation and paneling layout is
   a. locating the studs.
   b. arranging panels on edge side by side around the room.
   c. cleaning the wall and scraping down high spots.
   d. removing window, door, and baseboard trim.

_____ 15. When positioning a panel, make sure the panel is perfectly plumb and its outer edge is over the
   a. drywall seam.
   b. centerline of a stud.
   c. furring strip.
   d. middle of the drywall panel.

_____ 16. When installing paneling over masonry, use
   a. MDF panels.
   b. panels that are 5/32" thick.
   c. mastic, not nails.
   d. furring strips.

_____ 17. An advantage of installing paneling horizontally is that
   a. furring strips can be used.
   b. blocking is not required.
   c. scribing is not necessary.
   d. conditioning occurs more quickly.

_____ 18. Which of the following statements is correct about estimating the amount of sheet paneling needed?
   a. First, figure the area of the room to be paneled.
   b. Divide the perimeter of the room by the width of the panel.
   c. Multiply the perimeter of the room by the width of the panel.
   d. Subtract the 5 percent waste allowance from the area of the room.

_____ 19. Estimate the number of 4×8 sheet panels needed for a room measuring 12' × 16' with a ceiling height of 8'.
   a. 14
   b. 28
   c. 15
   d. 14.7

_____ 20. The maximum width recommended for board paneling in the United States is
   a. 8".
   b. 4".
   c. 2'.
   d. 10".

Mechanicals

MATCHING

Directions: Match each word or phrase to the statement or sentence.

Column I

_____ 1. Plumbing device that receives or drains water.

_____ 2. Pipe that brings municipal water to a house.

_____ 3. Cable or group of cables that supplies electricity to an area or appliance.

_____ 4. A measure of electrical current.

_____ 5. Fresh air leaking into a house through cracks.

_____ 6. In construction, the plumbing, electrical, and HVAC systems.

_____ 7. Curved section of drainpipe that prevents sewer gases from entering a house.

_____ 8. Network of perforated pipes that filter liquid wastes from a septic system.

Column II

a. mechanicals
b. infiltration
c. trap
d. drain field
e. fixture
f. receptacle
g. ampere
h. circuit
i. service main

MULTIPLE CHOICE

Directions: In the blank provided, write the letter of the answer that best completes the statement or answers the question.

_____ 9. The correct order of the path of a plumbing system through a city home is
   a. service main, meter, supply pipes, waste pipes.
   b. meter, supply pipes, service main, waste pipes.
   c. supply pipes, meter, service main, waste pipes.
   d. sewage treatment, service main, supply pipes, meter.

_____ 10. Main water distribution lines in a home are typically _?_ pipe.
   a. ½" or ¾" PVC
   b. 1 ½" or 2" copper
   c. ¾" or 1" copper
   d. 2" or 2 ½" PVC

(Continued on next page)

CHAPTER 38 TEST (Continued)

_____ 11. PVC and ABS pipes are often used for ? in homes, but cast-iron is the higher-quality choice.
   a. waste systems
   b. water supply systems
   c. branch distribution lines
   d. HVAC systems

_____ 12. In an electrical service panel, the ? should be located at the top.
   a. general switch
   b. appliance switch
   c. HVAC switch
   d. master switch

_____ 13. What type of wiring is used for large kitchen appliances?
   a. No. 12 wire and 20-amp circuit breaker
   b. No. 12 wire and 15-amp circuit breaker
   c. No. 14 wire and 15-amp circuit breaker
   d. No. 14 wire and 30-amp circuit breaker

_____ 14. Electrical installations are inspected after ? and after the finish stage.
   a. framing
   b. rough-in
   c. mechanicals installation
   d. foundation completion

_____ 15. The abbreviation HVAC stands for
   a. high-voltage alternating current.
   b. heating, ventilating, and air conditioning
   c. hydronic ventilating and conditioning.
   d. home ventilating and air conditioning.

_____ 16. Which statement is **not** considered an advantage of forced-air heating systems?
   a. responds quickly to outdoor temperature changes
   b. can be used in many types of houses
   c. systems are very quiet
   d. ducts can also be used for air conditioning

_____ 17. Radiant heating systems are comfortable and quiet, and use either electric or ? systems.
   a. heat-pump
   b. air-handler
   c. compressor
   d. hydronic

_____ 18. Split-system central air conditioners generally place the ? outside.
   a. evaporator coil
   b. condenser coil
   c. air handler
   d. ducting

_____ 19. A(n) ? transfers heat from stale air it is exhausting to fresh air it is drawing in.
   a. HVAC system
   b. whole-house ventilator
   c. heat recovery ventilator
   d. forced hot-air heating system

_____ 20. Split-system air conditioning works on the principle that the ? absorbs heat and changes from a liquid to a gas and back into a liquid.
   a. refrigerant
   b. water vapor
   c. coil system
   d. convector

Thermal and Acoustical Insulation

MATCHING

Directions: Match each word or phrase to the statement or sentence.

Column I

_____ 1. Travels in a straight line away from a hot surface and heats anything solid it meets.

_____ 2. A measure of a material's ability to resist heat transmission.

_____ 3. Conditioned space that separates heated from unheated spaces.

_____ 4. A measure based on decibels to describe noise caused by footsteps, dropped objects, or furniture being moved.

_____ 5. A material highly resistant to moisture transmission.

_____ 6. Must be low for radiant-barrier materials to perform properly.

_____ 7. Must be high for radiant-barrier materials to perform properly.

_____ 8. A measure of the resistance of a building element to the passage of airborne sound.

_____ 9. When sound passes around a material instead of through it.

_____ 10. A measure of water vapor transmission through a material.

Column II

a. reflectivity
b. perm value
c. radiant heat
d. vapor barrier
e. flanking path
f. R-value
g. Impact Insulation Class
h. Impact Noise Rating
i. emissivity
j. Sound Transmission Class
k. thermal envelope

(Continued on next page)

CHAPTER 39 TEST (Continued)

MULTIPLE CHOICE

Directions: In the blank provided, write the letter of the answer that best completes the statement or answers the question.

_____ 11. The most common type of insulation used in houses is _?_ insulation.
   a. loose-fill   b. rigid   c. flexible   d. spray-foam

_____ 12. The type of insulation that seals cavities better than other types of insulation is _?_ insulation.
   a. spray-foam   b. flexible   c. rigid   d. loose-fill

_____ 13. On attic floors where pipes and wiring make installing blanket insulation difficult, _?_ insulation is often used.
   a. rigid   b. spray-foam   c. flexible   d. loose-fill

_____ 14. In areas with colder climates, an excellent R-value for insulation intended for ceilings is
   a. R-3.   b. R-5.6.   c. R-38.   d. R-11.

_____ 15. Some types of _?_ insulation can be used below grade on the exterior of basement walls.
   a. flexible   b. rigid   c. loose-fill   d. spray-foam

_____ 16. To prevent damage to exterior paint, interior finishes, and structural members, install a _?_ barrier on the heated side of a wall or ceiling.
   a. sound-absorption   c. vapor
   b. radiant-heat   d. sound-insulation

_____ 17. Which of the following provides the most satisfactory sound-resistant wall at the most economical cost?
   a. interior masonry wall
   b. drywall or lath and plaster
   c. sound-deadening insulating board and gypsum board
   d. double wall

_____ 18. Besides removing moisture and lowering attic temperatures, ventilation in an attic or roof is important because it
   a. reduces the formation of ice at the eaves.
   b. allows higher emissivity.
   c. provides greater resistance to airborne sounds.
   d. helps reduce sound by providing a flanking path.

_____ 19. Batts, blankets, and unfaced insulation are forms of _?_ insulation.
   a. rigid   b. loose-fill   c. spray-foam   d. flexible

_____ 20. The best type of insulation to use around doors and windows is _?_ insulation.
   a. flexible   b. spray-foam   c. loose-fill   d. rigid sheet

Walls and Ceilings

MATCHING

Directions: Match each word or phrase to the statement or sentence.

Column I

_____ 1. A material that serves as a base for another material.

_____ 2. Can be nailed, screwed, or crimped to protect raw edges of drywall at a corner.

_____ 3. Smoothing, as when applying joint compound.

_____ 4. Used to coat gypsum-base drywall.

_____ 5. Base material for plaster.

_____ 6. A material permanently or temporarily attached to a surface to be plastered.

_____ 7. Consists of panels held in place by a metal or plaster grid at a distance from overhead floor joists.

_____ 8. Consists of panels glued directly to the ceiling surface.

_____ 9. Made from sand, lime, and water.

_____ 10. Green board.

_____ 11. Blue board.

Column II

a. Cornerite
b. feathering
c. substrate
d. plaster
e. veneer plaster
f. ground
g. acoustical ceiling
h. corner bead
i. MR drywall
j. suspended ceiling
k. gypsum base drywall
l. lath

(Continued on next page)

CHAPTER 40 TEST *(Continued)*

MULTIPLE CHOICE

Directions: In the blank provided, write the letter of the answer that best completes the statement or answers the question.

_____ 12. The type of drywall that would be the best choice to provide fire resistance to ceilings is _?_ drywall.
  a. MR
  b. Type-X fire-code
  c. Type-C fire-code
  d. fire-code MR

_____ 13. The type of drywall that would be used for a bathroom is _?_ drywall.
  a. fire-code MR
  b. standard
  c. Type-X fire-code
  d. Type-C fire-code

_____ 14. Drywall nails for ½" drywall should be at least _?_ long.
  a. 1"
  b. 1 ¼"
  c. 1 ½"
  d. 1 ⅜"

_____ 15. To reduce the problem of nail pops,
  a. use moisture-resistant drywall.
  b. apply the drywall over framing lumber with a high moisture content to allow for shrinkage.
  c. fasten drywall panels directly to the face of wide dimensional lumber.
  d. make sure the moisture content of the framing is less than 15 percent.

_____ 16. To reduce exposure to joint compound sanding dust, use
  a. stilts and a drywall router.
  b. a screw gun instead of a drywall hammer.
  c. topping compound instead of all-purpose joint compound.
  d. a pole sander or a vacuum-based sanding system.

_____ 17. The type of lath often used around tub recesses or other bath and kitchen areas is the _?_ lath.
  a. wood
  b. metal
  c. gypsum
  d. reinforcing

_____ 18. Which of the following is the correct order used in three-coat plaster work?
  a. scratch coat, brown coat, finish coat
  b. double-up work, finish coat
  c. brown coat, scratch coat, finish coat
  d. scratch coat, brown coat, leveling coat

_____ 19. The plaster finish that is commonly done where a gloss or enamel paint will be used is the _?_ finish.
  a. sand-float
  b. putty
  c. brown-coat
  d. double-up

_____ 20. When installing a suspended ceiling, use _?_ to mark the locations of the first main beam and the first row of cross tees.
  a. wires
  b. wall molding
  c. strings
  d. chalk lines

Exterior and Interior Finishes

MATCHING

Directions: Match each word or phrase to the statement or sentence.

Column I

_____ 1. Paint with more binder than standard paint.

_____ 2. Paint with binder suspended in water.

_____ 3. Synthetic binder used in oil-base paint.

_____ 4. Paint's gloss when it is dry.

_____ 5. The liquid part of paint that suspends pigment and binders.

_____ 6. Paint ingredient that adds color.

_____ 7. Painting in corners with a brush.

_____ 8. Resin that holds paint pigment together.

Column II

a. pigment
b. carrier
c. primer
d. edging
e. binder
f. acrylic
g. sheen
h. latex
i. grain
j. alkyd

MULTIPLE CHOICE

Directions: In the blank provided, write the letter of the answer that best completes the statement or answers the question.

_____ 9. The most common type of a film-forming finish is
  a. semi-transparent stain.
  b. paint.
  c. pigment.
  d. transparent stain.

_____ 10. Paint coats a wood's surface, while a penetrating finish such as _?_ soaks into the wood.
  a. a solvent
  b. linseed oil
  c. a oil-base semi-transparent stain
  d. a solid-color stain

_____ 11. Binders in latex paints are suspended in
  a. mineral spirits.
  b. turpentine.
  c. water.
  d. oils such as tung oil.

(Continued on next page)

CHAPTER 41 TEST (Continued)

_____ 12. In general, exterior paint lasts _?_ years before requiring recoating.
   a. three to seven
   b. ten to fifteen
   c. two to five
   d. seven to ten

_____ 13. When painting the exterior of a house with oil-base paint, best results are obtained when the temperature stays _?_ for 24 hours after application.
   a. above 40°F
   b. above 50°F
   c. below 40°F
   d. between 60°F and 70°F

_____ 14. To avoid drips and splatters ruining newly painted sections, you should paint from
   a. left to right.
   b. the top down.
   c. right to left.
   d. narrower to wider sections.

_____ 15. To get the best finished appearance with a brush, you should paint _?_ the grain.
   a. against
   b. with
   c. perpendicular to
   d. diagonally across

_____ 16. After painting a window, you should _?_ to keep it from sticking shut.
   a. leave it open
   b. put wedges under it
   c. move it up and down
   d. rub the edges lightly with a damp rag

_____ 17. If paint cracks or alligators, it's likely that it was applied in heavy coats without
   a. sufficient drying time between coats.
   b. primer being applied first.
   c. being stirred properly.
   d. the surface being adequately sanded.

_____ 18. When paint blisters or peels, it is usually because _?_ has/have pulled moisture from the wood.
   a. chemicals in the paint
   b. the curing paint
   c. the primer
   d. heat from the sun

_____ 19. To even out slight variations in color between cans of paint, _?_ the paint by mixing one can with the next.
   a. "meld"
   b. "feather"
   c. "box"
   d. "roll in"

_____ 20. When painting interior surfaces, protect edges that you don't want to paint with _?_ tape.
   a. standard masking
   b. painter's masking
   c. edging
   d. duct

Carpentry & Building Construction Instructor Resource Guide
Copyright © Glencoe/McGraw-Hill

Wood Flooring

MATCHING

Directions: Match each word or phrase to the statement or sentence.

Column I

_____ 1. Process used on tongue-and-groove flooring to conceal the fasteners.

_____ 2. Solid-wood board that is at least 3" wide.

_____ 3. Narrow length of wood no more than 3 ½" wide.

_____ 4. Wood flooring assembled into a geometric design.

_____ 5. Occurs when wood flooring reaches a moisture content equal to that in the place of installation.

_____ 6. Top layer of engineered-wood flooring.

_____ 7. Length of lumber used to support wood flooring over concrete.

_____ 8. Solvents used in floor finishes that present environmental concerns.

_____ 9. Modified urethane floor finishes.

_____ 10. The difference in the height of adjacent boards after installation.

Column II

a. strip
b. overwood
c. blind nailing
d. polyurethanes
e. plank
f. sleeper
g. shorts
h. wear layer
i. volatile organic compounds
j. acclimation
k. parquet

MULTIPLE CHOICE

Directions: In the blank provided, write the letter of the answer that best completes the statement or answers the question.

_____ 11. The most plentiful and the popular wood flooring in the United States is
    a. maple.
    b. oak.
    c. walnut.
    d. cherry.

(Continued on next page)

CHAPTER 42 TEST (Continued)

_____ 12. The four most common hardwoods used as a finish floor material are maple, oak, birch, and
   a. mahogany.
   b. walnut.
   c. cherry.
   d. beech.

_____ 13. Solid-wood flooring that is similar to engineered-wood flooring is
   a. parquet tile.
   b. bevel-edged planking.
   c. oak-strip flooring.
   d. No. 2 Common shorts.

_____ 14. If wood flooring absorbs moisture from the subfloor, the underside of the flooring expands and may cause
   a. boards to cup.
   b. nail pops.
   c. boards to shrink.
   d. glue to weaken.

_____ 15. How can a straight course be ensured when laying a strip floor over a wood subfloor?
   a. Line up the first course with a wall that is at right angles to the floor joists.
   b. Snap two chalk lines on diagonals from the corners to find the center of the room.
   c. Stretch a string the length of the room between two nails 8" from a side wall.
   d. Line up the first course with a wall that is parallel to the floor joists.

_____ 16. When installing a strip floor over a wood subfloor, what can be done to protect the flooring from moisture and help prevent squeaks?
   a. Correct defects on the subfloor and then sweep thoroughly.
   b. Sand the subfloor and caulk seams.
   c. Coat the subfloor with water-resistant paint.
   d. Staple building paper to the subfloor.

_____ 17. Before installing wood flooring over a concrete subfloor, allow the concrete to cure for
   a. two days.
   b. two weeks.
   c. two months.
   d. six months.

_____ 18. How should a vapor barrier be installed when laying wood flooring over sleepers?
   a. Cover the asphalt mastic with 4-mil polyethylene.
   b. Spread 6-mil polyethylene over the sleepers.
   c. Roll out strips of 15-lb. asphalt felt before laying sleepers.
   d. Coat the concrete with water-resistant paint and then put down sleepers.

_____ 19. The method of installing engineered-wood flooring that can be used to apply the flooring directly to a concrete subfloor is the _?_ method.
   a. glue-down
   b. floating
   c. closed-cell foam padding
   d. nail-down

_____ 20. When installing closing boards, use a _?_ to lever the boards into place.
   a. chisel
   b. prybar
   c. screwdriver
   d. hammer

Resilient Flooring and Ceramic Tile

MATCHING

Directions: Match each word or phrase to the statement or sentence.

Column I

_____ 1. A thin, smooth-surfaced panel that prevents subfloor flaws from showing through resilient flooring.

_____ 2. A ceramic tile without the glaze.

_____ 3. Thin mortar used for filling spaces between ceramic tiles.

_____ 4. Hand tool used to cut through the fiberglass reinforcement of cement board.

_____ 5. Tool used to cut tiles in a straight line.

_____ 6. The ability of a substance to allow water to flow through it.

_____ 7. Any tile that is 2" square or smaller.

_____ 8. A pliers-like tool used to cut shapes in ceramic tile.

_____ 9. A glassy finish for ceramic tile.

_____ 10. A very effective adhesive made of Portland cement, sand, and additives.

_____ 11. Cement-based sheets that provide an excellent base for tile.

_____ 12. Tiles at least ½" thick that are used for floors.

Column II

a. glaze
b. backerboard
c. permeability
d. dry-set mortar
e. bisque
f. nippers
g. hand-held snap cutter
h. scoring tool
i. grout
j. mosaic tile
k. paver tiles
l. resilient flooring
m. underlayment

(Continued on next page)

CHAPTER 43 TEST *(Continued)*

MULTIPLE CHOICE

Directions: In the blank provided, write the letter of the answer that best completes the statement or answers the question.

_____ 13. Sheet vinyl comes in large rolls that are _?_ wide.
   a. 4'
   b. 6'
   c. 12'
   d. 16'

_____ 14. Which tool works best for cutting vinyl flooring?
   a. nippers
   b. utility knife
   c. snap cutter
   d. sewing scissors

_____ 15. A room 10' by 12' would require approximately _?_ sq. ft. of sheet vinyl.
   a. 100
   b. 120
   c. 130
   d. 137

_____ 16. How many 12" vinyl tiles would be required to cover a floor that is 15' by 12'?
   a. 100
   b. 120
   c. 180
   d. 240

_____ 17. The best tool to use for spreading tile adhesive is the
   a. grout saw.
   b. scoring tool.
   c. trowel.
   d. utility knife.

_____ 18. The best tool to use for cleaning cured grout from between tiles is the
   a. grout saw.
   b. scoring tool.
   c. trowel.
   d. nibbler.

_____ 19. The best permeability for ceramic tiles that are exposed to water in a bathroom is
   a. impervious.
   b. vitreous.
   c. semi-vitreous.
   d. nonvitreous.

_____ 20. Backerboard provides a particularly excellent base for ceramic tile installed on a(n)
   a. patio.
   b. entryway.
   c. kitchen floor.
   d. shower stall floor.

Chimneys and Fireplaces

MATCHING

Directions: Match each word or phrase to the statement or sentence.

Column I

_____ 1. Where the burning takes place in a fireplace.

_____ 2. A course of brick that is offset to extend past the course below it.

_____ 3. Hinged lid used to vary the size of a chimney's throat opening.

_____ 4. Upward movement of air in a chimney.

_____ 5. Floor of a fireplace firebox, plus the area in front of it.

_____ 6. Part that keeps downdrafts from driving smoke back down into a firebox.

_____ 7. Passage inside a chimney through which air, gases, and smoke rise.

_____ 8. This replaces air that is exhausted by a combustion appliance.

_____ 9. The front of a fireplace.

_____ 10. Steel angle iron across the top of the firebox opening to support the masonry.

_____ 11. Area from the top of the fireplace throat to the bottom of the flue.

_____ 12. Narrowest part of a fireplace firebox, where the damper is located.

Column II

a. flue
b. lintel
c. corbel
d. draft
e. makeup air
f. mantel
g. throat
h. firebox
i. smoke shelf
j. hearth
k. surround
l. damper
m. smoke chamber

(Continued on next page)

CHAPTER 44 TEST (Continued)

MULTIPLE CHOICE

Directions: In the blank provided, write the letter of the answer that best completes the statement or answers the question.

_____ 13. Any portion of a chimney located within the house must have a _?_ clearance between its walls and the wood framing.
   a. 1"
   b. 2"
   c. 3"
   d. 4"

_____ 14. Which of these characteristics of a chimney flue does **not** affect how well the chimney draws?
   a. height
   b. material
   c. shape
   d. smoothness

_____ 15. When completing the top of a chimney, what is the preferred, safest surface to work from?
   a. a crane
   b. an extension ladder
   c. scaffolding
   d. the roof

_____ 16. For a fireplace with an opening less than 6 square feet in area, the front hearth should extend out at least _?_ in front and 8" on both sides.
   a. 6"
   b. 10"
   c. 12"
   d. 16"

_____ 17. A mortar or concrete cap is placed over the top course of chimney brick
   a. to keep birds and squirrels from nesting inside the chimney.
   b. to stop sparks from escaping.
   c. to hold the top course of chimney bricks in place.
   d. to prevent moisture from seeping between the brick and flue liner.

_____ 18. Since the chimney is usually the heaviest part of a building, the cement chimney foundation must be at least _?_ thick.
   a. 3"
   b. 6"
   c. 12"
   d. 18"

_____ 19. Due to the fire hazard, no woodwork can be placed within _?_ of a firebox opening.
   a. 6"
   b. 12"
   c. 18"
   d. 24"

_____ 20. What type of fireplace can be placed directly against wood framing?
   a. non-combustible
   b. zero-clearance
   c. freestanding
   d. corbel

Decks and Porches

MATCHING

Directions: Match each word or phrase to the statement or sentence.

Column I

_____ 1. The least decay-resistant part of the tree.

_____ 2. Portion of the tree nearest to the core.

_____ 3. A wood preservative that could contaminate soil and ground water.

_____ 4. A preservative used in wood treatment that does not contain arsenic.

_____ 5. A synthetic product made of wood dust or fibers and recycled plastic.

_____ 6. Process that coats steel with a protective layer of zinc.

_____ 7. Concrete column that supports a concentrated load, such as a post.

_____ 8. A cylinder made of laminated, waxed paper products that serves as a form for concrete.

_____ 9. The length of lumber that connects the deck to the house.

_____ 10. An enlarged landing at the top of steps.

Column II

a. composite decking
b. stoop
c. plastic decking
d. sapwood
e. pier
f. ledger
g. CCA
h. galvanizing
i. heartwood
j. Sonotube
k. ACQ

MULTIPLE CHOICE

Directions: In the blank provided, write the letter of the answer that best completes the statement or answers the question.

_____ 11. The most popular types of decking are made from
   a. hardwood.
   b. softwood.
   c. plastic.
   d. composites.

_____ 12. The basic element of a deck that is secured to the pier by a metal anchor is the
   a. post.
   b. beam.
   c. joist.
   d. ledger.

(Continued on next page)

CHAPTER 45 TEST (Continued)

_____ 13. To determine the layout for piers for a rectangular attached deck,
   a. use a 12' 2×4 and a lever.
   b. measure 12' from the foundation wall in two places.
   c. use a block of wood and a plumb line.
   d. use the 3-4-5 method to ensure a right-angle layout.

_____ 14. To plumb a post, use
   a. the 3-4-5 method to figure a right angle.
   b. a tape measure and a shovel.
   c. a block of wood, a plumb line, and a tape measure.
   d. a plumb line and mark the distances between posts.

_____ 15. When cutting preservative-treated wood, always ? to prevent chemicals from leaching into the soil.
   a. treat the wood ahead of time with CCA
   b. wear gloves and a dust mask
   c. cut over a tarp
   d. flush the cut with running water

_____ 16. Concrete porch steps that are precast are
   a. hollow to reduce their weight.
   b. cast in a factory and allowed to cure on site.
   c. constructed on site by masons.
   d. partly made from wood or metal.

_____ 17. The basic elements of a deck are
   a. softwoods, hardwoods, plastics, and composites.
   b. redwood, cedar, mahogany, and teak.
   c. appearance, strength, and moisture content.
   d. piers, ledgers, posts, railings, and beams.

_____ 18. The type of decking material that is hollow or partially hollow is
   a. composite.   c. sapwood.
   b. plastic.   d. heartwood.

_____ 19. When compared to a solid wood beam, a built-up beam has the advantage of being
   a. less likely to rot.   c. stronger.
   b. easier to position.   d. as deep as it is wide.

_____ 20. To avoid having fasteners that show, many builders use ? to fasten the decking to the joists.
   a. glue   c. face nailing
   b. toenailing   d. blind nailing

Answer Key for Chapter Tests

■ Chapter 1 Test Answers
1. f Section 1.2
2. c Section 1.5
3. e Section 1.1
4. h Section 1.1
5. d Section 1.3
6. a Section 1.6
7. b Section 1.3
8. i Section 1.3
9. b Section 1.1
10. d Section 1.1
11. a Section 1.2
12. b Section 1.4
13. b Section 1.4
14. d Section 1.5
15. b Section 1.5
16. c Section 1.5
17. a Section 1.5
18. b Section 1.6
19. d Section 1.6
20. b Section 1.6

■ Chapter 2 Test Answers
1. j Section 2.1
2. l Section 2.1
3. f Section 2.1
4. c Section 2.2
5. a Section 2.2
6. i Section 2.2
7. b Section 2.2
8. h Section 2.2
9. e Section 2.1
10. d Section 2.1
11. d Section 2.1
12. b Section 2.2
13. d Section 2.2
14. a Section 2.2
15. d Section 2.2
16. c Section 2.2
17. b Section 2.2
18. a Section 2.2
19. c Section 2.2
20. b Section 2.2

■ Chapter 3 Test Answers
1. b Section 3.1
2. f Section 3.3
3. h Section 3.2
4. d Section 3.3
5. j Section 3.3
6. c Section 3.3
7. g Section 3.3
8. i Section 3.3
9. a Section 3.3
10. e Section 3.1
11. c Section 3.1
12. b Section 3.2
13. a Section 3.2
14. d Section 3.2
15. c Section 3.2
16. b Section 3.2
17. b Section 3.2
18. d Section 3.3
19. a Section 3.3
20. c Section 3.3

■ Chapter 4 Test Answers
1. d Section 4.1
2. a Section 4.1
3. e Section 4.1
4. f Section 4.1
5. b Section 4.1
6. g Section 4.2
7. b Section 4.2
8. c Section 4.1
9. c Section 4.1
10. d Section 4.1
11. c Section 4.1
12. a Section 4.1
13. c Section 4.1
14. b Section 4.1
15. d Section 4.2
16. c Section 4.2
17. c Section 4.2
18. b Section 4.2
19. a Section 4.2
20. d Section 4.2

■ Chapter 5 Test Answers
1. f Section 5.3
2. b Section 5.2
3. a Section 5.3
4. i Section 5.3
5. h Section 5.3
6. c Section 5.3
7. j Section 5.3
8. e Section 5.3
9. d Section 5.3
10. e Section 5.3
11. i Section 5.3
12. f Section 5.3
13. c Section 5.3
14. g Section 5.3
15. d Section 5.3
16. j Section 5.3
17. a Section 5.3
18. b Section 5.3
19. d Section 5.1
20. c Section 5.2
21. b Section 5.2
22. a Section 5.3
23. c Section 5.3
24. b Section 5.2
25. b Section 5.3
26. d Section 5.3
27. a Section 5.3
28. c Section 5.3
29. b Section 5.3
30. c Section 5.1
31. d Section 5.3
32. b Section 5.3
33. d Section 5.1

■ Chapter 6 Test Answers
1. d Section 6.1
2. a Section 6.1
3. c Section 6.2
4. b Section 6.2
5. e Section 6.2
6. e Section 6.1
7. b Section 6.1
8. d Section 6.1
9. a Section 6.3
10. c Section 6.3
11. b Section 6.1
12. a Section 6.2
13. c Section 6.2
14. c Section 6.2
15. a Section 6.4
16. b Section 6.5
17. d Section 6.5
18. d Section 6.5
19. c Section 6.5
20. b Section 6.2

■ Chapter 7 Test Answers
1. b Section 7.1
2. f Section 7.1
3. a Section 7.1
4. e Section 7.5
5. c Section 7.1
6. b Section 7.6
7. c Section 7.6
8. a Section 7.6
9. d Section 7.6
10. d Section 7.6
11. a Section 7.1
12. c Section 7.1
13. b Section 7.1
14. b Section 7.2
15. b Section 7.2
16. b Section 7.2
17. d Section 7.2
18. c Section 7.4

ANSWER KEY FOR CHAPTER TESTS (Continued)

19. c   Section 7.5
20. c   Section 7.5

Chapter 8 Test Answers
1. e   Section 8.1
2. f   Section 8.1
3. a   Section 8.1
4. d   Section 8.1
5. c   Section 8.1
6. b   Section 8.1
7. f   Section 8.1
8. c   Section 8.1
9. e   Section 8.1
10. d   Section 8.1
11. b   Section 8.1
12. a   Section 8.1
13. b   Section 8.1
14. c   Section 8.1
15. a   Section 8.1
16. d   Section 8.1
17. b   Section 8.1
18. c   Section 8.2
19. b   Section 8.2
20. d   Section 8.2
21. a   Section 8.2
22. b   Section 8.2
23. c   Section 8.2
24. a   Section 8.2
25. d   Section 8.2

Chapter 9 Test Answers
1. c   Section 9.3
2. g   Section 9.3
3. b   Section 9.1
4. a   Section 9.2
5. f   Section 9.2
6. e   Section 9.4
7. e   Section 9.3
8. c   Section 9.3
9. d   Section 9.3
10. b   Section 9.1
11. g   Section 9.2
12. f   Section 9.1
13. c   Section 9.1
14. d   Section 9.1
15. a   Section 9.1
16. c   Section 9.1
17. b   Section 9.1
18. c   Section 9.2
19. d   Section 9.2
20. a   Section 9.2
21. b   Section 9.2
22. d   Section 9.2
23. a   Section 9.3
24. b   Section 9.3
25. a   Section 9.3
26. d   Section 9.3
27. a   Section 9.4
28. c   Section 9.4

29. d   Section 9.3
30. d   Section 9.1
31. b   Section 9.3
32. c   Section 9.1
33. b   Section 9.3

Chapter 10 Test Answers
1. f   Section 10.2
2. c   Section 10.2
3. d   Section 10.2
4. a   Section 10.2
5. g   Section 10.2
6. b   Section 10.2
7. d   Section 10.1
8. b   Section 10.1
9. a   Section 10.1
10. c   Section 10.1
11. f   Section 10.1
12. c   Section 10.1
13. d   Section 10.1
14. b   Section 10.1
15. c   Section 10.1
16. c   Section 10.2
17. a   Section 10.2
18. d   Section 10.2
19. a   Section 10.2
20. a   Section 10.1

Chapter 11 Test Answers
1. e   Section 11.1
2. g   Section 11.1
3. i   Section 11.1
4. a   Section 11.2
5. n   Section 11.2
6. l   Section 11.2
7. f   Section 11.1
8. b   Section 11.2
9. m   Section 11.1
10. c   Section 11.1
11. h   Section 11.1
12. d   Section 11.1
13. j   Section 11.1
14. b   Section 11.1
15. c   Section 11.1
16. a   Section 11.2
17. b   Section 11.2
18. d   Section 11.1
19. b   Section 11.1
20. c   Section 11.1

Chapter 12 Test Answers
1. d   Section 12.1
2. b   Section 12.1
3. c   Section 12.1
4. a   Section 12.1
5. f   Section 12.1
6. e   Section 12.1
7. a   Section 12.1
8. c   Section 12.1

9. f   Section 12.1
10. d   Section 12.1
11. c   Section 12.1
12. d   Section 12.1
13. d   Section 12.1
14. c   Section 12.1
15. c   Section 12.1
16. a   Section 12.1
17. b   Section 12.2
18. b   Section 12.2
19. a   Section 12.2
20. c   Section 12.2

Chapter 13 Test Answers
1. c   Section 13.2
2. f   Section 13.1
3. i   Section 13.1
4. k   Section 13.1
5. a   Section 13.1
6. j   Section 13.1
7. e   Section 13.1
8. h   Section 13.1
9. g   Section 13.1
10. d   Section 13.1
11. b   Section 13.1
12. c   Section 13.1
13. c   Section 13.1
14. a   Section 13.1
15. c   Section 13.1
16. a   Section 13.2
17. d   Section 13.2
18. d   Section 13.2
19. b   Section 13.2
20. b   Section 13.1

Chapter 14 Test Answers
1. g   Section 14.2
2. c   Section 14.3
3. d   Section 14.3
4. b   Section 14.3
5. f   Section 14.3
6. a   Section 14.2
7. c   Section 14.1
8. e   Section 14.1
9. a   Section 14.1
10. d   Section 14.1
11. f   Section 14.2
12. b   Section 14.1
13. c   Section 14.1
14. d   Section 14.1
15. b   Section 14.1
16. a   Section 14.1
17. c   Section 14.1
18. d   Section 14.2
19. d   Section 14.2
20. b   Section 14.2
21. a   Section 14.2
22. c   Section 14.2
23. c   Section 14.2

Carpentry & Building Construction Instructor Resource Guide
Copyright © Glencoe/McGraw-Hill

ANSWER KEY FOR CHAPTER TESTS (Continued)

24. b   Section 14.2
25. b   Section 14.2
26. c   Section 14.2
27. d   Section 14.3
28. b   Section 14.3
29. c   Section 14.3
30. d   Section 14.3
31. a   Section 14.3
32. a   Section 14.3
33. b   Section 14.3

■ Chapter 15 Test Answers

1. b   Section 15.1
2. k   Section 15.1
3. i   Section 15.1
4. g   Section 15.2
5. c   Section 15.2
6. j   Section 15.1
7. f   Section 15.1
8. h   Section 15.1
9. a   Section 15.1
10. d   Section 15.2
11. b   Section 15.1
12. a   Section 15.1
13. d   Section 15.1
14. a   Section 15.2
15. d   Section 15.2
16. b   Section 15.2
17. c   Section 15.1
18. a   Section 15.1
19. a   Section 15.1
20. c   Section 15.1

■ Chapter 16 Test Answers

1. c   Section 16.1
2. f   Section 16.1
3. h   Section 16.1
4. k   Section 16.1
5. i   Section 16.1
6. a   Section 16.1
7. d   Section 16.1
8. g   Section 16.1
9. e   Section 16.1
10. b   Section 16.1
11. a   Section 16.1
12. c   Section 16.1
13. a   Section 16.2
14. c   Section 16.2
15. a   Section 16.2
16. b   Section 16.2
17. c   Section 16.2
18. b   Section 16.1
19. c   Section 16.1
20. a   Section 16.1

■ Chapter 17 Test Answers

1. b   Section 17.2
2. e   Section 17.3
3. h   Section 17.1
4. l   Section 17.2
5. d   Section 17.1
6. j   Section 17.1
7. g   Section 17.4
8. i   Section 17.1
9. f   Section 17.2
10. c   Section 17.1
11. b   Section 17.1
12. a   Section 17.1
13. d   Section 17.3
14. c   Section 17.1
15. b   Section 17.1
16. a   Section 17.3
17. d   Section 17.4
18. a   Section 17.4
19. d   Section 17.2
20. b   Section 17.4

■ Chapter 18 Test Answers

1. b   Section 18.2
2. a   Section 18.1
3. f   Section 18.2
4. d   Section 18.2
5. c   Section 18.2
6. c   Section 18.1
7. e   Section 18.1
8. a   Section 18.1
9. b   Section 18.1
10. d   Section 18.1
11. a   Section 18.1
12. c   Section 18.1
13. b   Section 18.2
14. d   Section 18.1
15. c   Section 18.1
16. a   Section 18.1
17. d   Section 18.1
18. a   Section 18.1
19. c   Section 18.2
20. b   Section 18.2

■ Chapter 19 Test Answers

1. c   Section 19.1
2. g   Section 19.1
3. i   Section 19.1
4. d   Section 19.3
5. j   Section 19.1
6. f   Section 19.3
7. k   Section 19.1
8. a   Section 19.1
9. h   Section 19.2
10. e   Section 19.2
11. b   Section 19.1
12. c   Section 19.1
13. d   Section 19.1
14. a   Section 19.1
15. c   Section 19.1
16. d   Section 19.1
17. b   Section 19.2
18. b   Section 19.2
19. a   Section 19.3
20. d   Section 19.2

■ Chapter 20 Test Answers

1. i   Section 20.4
2. h   Section 20.1
3. c   Section 20.1
4. e   Section 20.1
5. k   Section 20.2
6. j   Section 20.2
7. b   Section 20.2
8. d   Section 20.2
9. a   Section 20.2
10. g   Section 20.2
11. c   Section 20.1
12. a   Section 20.2
13. d   Section 20.2
14. c   Section 20.3
15. a   Section 20.2
16. b   Section 20.3
17. c   Section 20.4
18. b   Section 20.4
19. d   Section 20.2
20. a   Section 20.2
21. b   Section 20.3
22. a   Section 20.1
23. b   Section 20.2
24. d   Section 20.3
25. c   Section 20.1

■ Chapter 21 Test Answers

1. c   Section 21.1
2. f   Section 21.1
3. b   Section 21.1
4. a   Section 21.1
5. e   Section 21.1
6. g   Section 21.1
7. h   Section 21.1
8. f   Section 21.4
9. a   Section 21.4
10. b   Section 21.4
11. g   Section 21.4
12. e   Section 21.4
13. c   Section 21.4
14. c   Section 21.2
15. d   Section 21.2
16. a   Section 21.2
17. b   Section 21.2
18. d   Section 21.2
19. c   Section 21.2
20. c   Section 21.2
21. b   Section 21.3
22. d   Section 21.3
23. b   Section 21.1
24. a   Section 21.1
25. c   Section 21.5
26. d   Section 21.5
27. c   Section 21.3
28. a   Section 21.3

ANSWER KEY FOR CHAPTER TESTS (Continued)

29. d Section 21.3
30. c Section 21.3
31. b Section 21.1
32. c Section 21.5
33. d Section 21.3

■ Chapter 22 Test Answers

1. e Section 22.1
2. a Section 22.1
3. c Section 22.1
4. g Section 22.1
5. b Section 22.1
6. f Section 22.1
7. d Section 22.1
8. j Section 22.1
9. f Section 22.2
10. i Section 22.1
11. h Section 22.1
12. d Section 22.1
13. b Section 22.1
14. g Section 22.1
15. a Section 22.1
16. e Section 22.1
17. a Section 22.2
18. b Section 22.1
19. c Section 22.1
20. d Section 22.1
21. a Section 22.1
22. b Section 22.1
23. d Section 22.2
24. b Section 22.3
25. a Section 22.3

■ Chapter 23 Test Answers

1. d Section 23.1
2. h Section 23.1
3. k Section 23.3
4. e Section 23.1
5. a Section 23.1
6. g Section 23.1
7. j Section 23.2
8. b Section 23.2
9. f Section 23.1
10. c Section 23.1
11. d Section 23.1
12. b Section 23.1
13. a Section 23.1
14. c Section 23.1
15. b Section 23.2
16. c Section 23.2
17. d Section 23.1
18. c Section 23.1
19. a Section 23.3
20. c Section 23.2
21. a Section 23.3
22. b Section 23.3
23. b Section 23.3
24. a Section 23.3
25. c Section 23.2

■ Chapter 24 Test Answers

1. d Section 24.1
2. h Section 24.1
3. p Section 24.3
4. l Section 24.3
5. c Section 24.3
6. k Section 24.1
7. o Section 24.1
8. g Section 24.3
9. b Section 24.3
10. j Section 24.4
11. n Section 24.4
12. f Section 24.4
13. a Section 24.2
14. i Section 24.3
15. e Section 24.4
16. a Section 24.1
17. d Section 24.1
18. c Section 24.1
19. c Section 24.1
20. a Section 24.2
21. b Section 24.3
22. d Section 24.3
23. b Section 24.3
24. a Section 24.3
25. c Section 24.4
26. d Section 24.4
27. a Section 24.4
28. b Section 24.2
29. c Section 24.2
30. b Section 24.4
31. b Section 24.4
32. c Section 24.2
33. a Section 24.1

■ Chapter 25 Test Answers

1. b Section 25.2
2. d Section 25.1
3. g Section 25.1
4. j Section 25.1
5. c Section 25.1
6. f Section 25.1
7. i Section 25.1
8. k Section 25.1
9. e Section 25.1
10. a Section 25.1
11. b Section 25.1
12. c Section 25.1
13. b Section 25.1
14. a Section 25.1
15. d Section 25.1
16. c Section 25.2
17. a Section 25.2
18. b Section 25.2
19. d Section 25.2
20. b Section 25.2

■ Chapter 26 Test Answers

1. f Section 26.2
2. e Section 26.2
3. b Section 26.2
4. a Section 26.2
5. d Section 26.2
6. c Section 26.2
7. h Section 26.2
8. b Section 26.3
9. c Section 26.3
10. a Section 26.3
11. c Section 26.1
12. d Section 26.1
13. c Section 26.1
14. a Section 26.1
15. b Section 26.1
16. f Section 26.1
17. a Section 26.1
18. c Section 26.2
19. d Section 26.3
20. b Section 26.3
21. b Section 26.2
22. d Section 26.1
23. a Section 26.1
24. a Section 26.3
25. c Section 26.3

■ Chapter 27 Test Answers

1. c Section 27.1
2. g Section 27.3
3. a Section 27.1
4. f Section 27.1
5. e Section 27.1
6. c Section 27.1
7. b Section 27.3
8. b Section 27.1
9. d Section 27.1
10. a Section 27.1
11. d Section 27.1
12. b Section 27.1
13. a Section 27.1
14. d Section 27.1
15. c Section 27.2
16. c Section 27.2
17. b Section 27.2
18. d Section 27.2
19. b Section 27.2
20. a Section 27.2
21. c Section 27.2
22. a Section 27.2
23. d Section 27.2
24. b Section 27.3
25. c Section 27.3

■ Chapter 28 Test Answers

1. d Section 28.2
2. i Section 28.2
3. l Section 28.1

Carpentry & Building Construction Instructor Resource Guide
Copyright © Glencoe/McGraw-Hill

ANSWER KEY FOR CHAPTER TESTS (Continued)

4. g Section 28.1
5. a Section 28.1
6. e Section 28.1
7. h Section 28.1
8. m Section 28.1
9. b Section 28.2
10. n Section 28.1
11. k Section 28.1
12. f Section 28.1
13. c Section 28.1
14. e Section 28.1
15. d Section 28.1
16. a Section 28.1
17. b Section 28.1
18. c Section 28.1
19. d Section 28.1
20. c Section 28.1
21. d Section 28.1
22. c Section 28.2
23. b Section 28.2
24. c Section 28.2
25. c Section 28.1

■ Chapter 29 Test Answers

1. a Section 29.1
2. f Section 29.1
3. i Section 29.1
4. l Section 29.1
5. n Section 29.1
6. c Section 29.1
7. k Section 29.1
8. m Section 29.1
9. o Section 29.1
10. q Section 29.2
11. g Section 29.2
12. b Section 29.2
13. e Section 29.3
14. j Section 29.1
15. h Section 29.3
16. b Section 29.3
17. d Section 29.1
18. a Section 29.1
19. b Section 29.1
20. c Section 29.1
21. a Section 29.2
22. d Section 29.2
23. d Section 29.2
24. b Section 29.2
25. c Section 29.3
26. c Section 29.2
27. c Section 29.3
28. a Section 29.3
29. a Section 29.1
30. d Section 29.2
31. d Section 29.2
32. b Section 29.3
33. c Section 29.3

■ Chapter 30 Test Answers

1. d Section 30.1
2. f Section 30.1
3. h Section 30.1
4. m Section 30.1
5. p Section 30.1
6. l Section 30.1
7. o Section 30.2
8. a Section 30.2
9. n Section 30.2
10. k Section 30.1
11. i Section 30.1
12. e Section 30.1
13. b Section 30.3
14. g Section 30.2
15. j Section 30.2
16. c Section 30.1
17. d Section 30.1
18. a Section 30.1
19. c Section 30.1
20. d Section 30.3
21. a Section 30.3
22. b Section 30.3
23. c Section 30.4
24. c Section 30.4
25. a Section 30.2
26. b Section 30.2
27. d Section 30.4
28. b Section 30.3
29. c Section 30.3
30. a Section 30.4
31. c Section 30.4
32. b Section 30.1
33. c Section 30.1

■ Chapter 31 Test Answers

1. e Section 31.1
2. b Section 31.2
3. g Introduction
4. i Section 31.1
5. c Section 31.2
6. f Section 31.1
7. j Section 31.2
8. h Section 31.1
9. k Section 31.3
10. d Section 31.1
11. b Section 31.1
12. d Section 31.1
13. a Section 31.1
14. c Section 31.1
15. a Section 31.1
16. d Section 31.2
17. d Section 31.3
18. b Section 31.3
19. c Section 31.3
20. b Section 31.2

■ Chapter 32 Test Answers

1. c Section 32.1
2. f Section 32.1
3. j Section 32.1
4. n Section 32.1
5. q Section 32.2
6. i Section 32.2
7. p Section 32.2
8. e Section 32.2
9. m Section 32.2
10. a Section 32.2
11. k Section 32.3
12. g Section 32.4
13. b Section 32.4
14. o Section 32.4
15. l Section 32.1
16. h Section 32.1
17. c Section 32.1
18. d Section 32.1
19. c Section 32.1
20. a Section 32.2
21. c Section 32.3
22. d Section 32.3
23. d Section 32.4
24. b Section 32.4
25. c Section 32.4
26. d Section 32.4
27. a Section 32.1
28. a Section 32.2
29. d Section 32.3
30. a Section 32.4
31. d Section 32.4
32. c Section 32.3
33. a Section 32.4

■ Chapter 33 Test Answers

1. b Section 33.1
2. f Section 33.1
3. i Section 33.1
4. k Section 33.1
5. d Section 33.1
6. a Section 33.1
7. j Section 33.2
8. h Section 33.2
9. e Section 33.1
10. c Section 33.2
11. d Section 33.1
12. b Section 33.1
13. b Section 33.1
14. a Section 33.1
15. d Section 33.1
16. a Section 33.1
17. c Section 33.1
18. b Section 33.2
19. c Section 33.2
20. b Section 33.2

ANSWER KEY FOR CHAPTER TESTS (Continued)

■ Chapter 34 Test Answers
1. e   Section 34.1
2. b   Section 34.2
3. d   Section 34.1
4. a   Section 34.1
5. f   Section 34.1
6. g   Section 34.1
7. c   Section 34.3
8. h   Section 34.1
9. d   Section 34.3
10. a   Section 34.1
11. g   Section 34.2
12. f   Section 34.2
13. c   Section 34.1
14. b   Section 34.2
15. e   Section 34.1
16. h   Section 34.2
17. c   Section 34.1
18. d   Section 34.1
19. a   Section 34.1
20. c   Section 34.1
21. a   Section 34.3
22. d   Section 34.3
23. d   Introduction
24. b   Section 34.3
25. c   Section 34.3
26. a   Section 34.1
27. b   Section 34.3
28. c   Section 34.3
29. b   Section 34.3
30. d   Section 34.1
31. a   Section 34.3
32. b   Section 34.3
33. c   Section 34.3

■ Chapter 35 Test Answers
1. c   Section 35.1
2. k   Section 35.1
3. i   Section 35.2
4. a   Section 35.3
5. h   Section 35.3
6. j   Section 35.3
7. l   Section 35.3
8. b   Section 35.2
9. e   Section 35.3
10. d   Section 35.3
11. g   Section 35.3
12. a   Section 35.2
13. d   Section 35.3
14. b   Section 35.2
15. f   Section 35.3
16. c   Section 35.3
17. b   Section 35.1
18. b   Section 35.3
19. c   Section 35.3
20. d   Section 35.3
21. a   Section 35.3
22. d   Section 35.1
23. d   Section 35.1
24. c   Section 35.2
25. a   Section 35.2

■ Chapter 36 Test Answers
1. h   Section 36.1
2. f   Section 36.2
3. e   Section 36.1
4. b   Section 36.1
5. a   Section 36.1
6. k   Section 36.1
7. g   Section 36.1
8. c   Section 36.1
9. i   Section 36.1
10. j   Section 36.2
11. c   Section 36.1
12. d   Section 36.2
13. a   Section 36.2
14. b   Section 36.2
15. c   Section 36.2
16. a   Section 36.2
17. d   Section 36.2
18. b   Section 36.2
19. d   Section 36.2
20. c   Section 36.2
21. c   Section 36.2
22. b   Section 36.2
23. c   Section 36.2
24. a   Section 36.2
25. d   Section 36.2
26. c   Section 36.2
27. a   Section 36.3
28. d   Section 36.3
29. d   Section 36.3
30. c   Section 36.3
31. b   Section 36.3
32. c   Section 36.3
33. c   Section 36.3

■ Chapter 37 Test Answers
1. d   Section 37.1
2. i   Section 37.1
3. a   Section 37.1
4. j   Section 37.1
5. e   Section 37.1
6. l   Section 37.2
7. g   Section 37.2
8. b   Section 37.2
9. h   Section 37.3
10. k   Section 37.1
11. b   Section 37.1
12. a   Section 37.1
13. d   Section 37.1
14. c   Section 37.2
15. b   Section 37.2
16. d   Section 37.2
17. b   Section 37.3
18. b   Section 37.3
19. c   Section 37.3
20. a   Section 37.3

■ Chapter 38 Test Answers
1. e   Section 38.1
2. i   Section 38.1
3. h   Section 38.2
4. g   Section 38.2
5. b   Section 38.3
6. a   Introduction
7. c   Section 38.1
8. d   Section 38.1
9. a   Section 38.1
10. c   Section 38.1
11. a   Section 38.1
12. d   Section 38.2
13. a   Section 38.2
14. b   Section 38.2
15. b   Section 38.3
16. c   Section 38.3
17. d   Section 38.3
18. b   Section 38.3
19. c   Section 38.3
20. a   Section 38.3

■ Chapter 39 Test Answers
1. c   Section 39.1
2. f   Section 39.1
3. k   Section 39.1
4. h   Section 39.2
5. d   Section 39.1
6. i   Section 39.1
7. a   Section 39.1
8. j   Section 39.2
9. e   Section 39.2
10. b   Section 39.1
11. c   Section 39.1
12. a   Section 39.1
13. d   Section 39.1
14. c   Section 39.1
15. b   Section 39.1
16. c   Section 39.1
17. c   Section 39.2
18. a   Section 39.1
19. d   Section 39.1
20. b   Section 39.1

■ Chapter 40 Test Answers
1. c   Section 40.1
2. h   Section 40.1
3. b   Section 40.1
4. e   Section 40.1
5. l   Section 40.2
6. f   Section 40.2
7. j   Section 40.3
8. g   Section 40.3
9. d   Section 40.2
10. i   Section 40.1
11. k   Section 40.2
12. c   Section 40.1
13. a   Section 40.1
14. b   Section 40.1

ANSWER KEY FOR CHAPTER TESTS (Continued)

15. d   Section 40.1
16. d   Section 40.1
17. b   Section 40.2
18. a   Section 40.2
19. b   Section 40.2
20. c   Section 40.3

■ **Chapter 41 Test Answers**

1. c   Section 41.1
2. h   Section 41.1
3. j   Section 41.1
4. g   Section 41.3
5. b   Section 41.1
6. a   Section 41.1
7. d   Section 41.3
8. e   Section 41.1
9. b   Section 41.1
10. c   Section 41.1
11. c   Section 41.1
12. d   Section 41.2
13. a   Section 41.2
14. b   Section 41.2
15. b   Section 41.2
16. c   Section 41.2
17. a   Section 41.2
18. d   Section 41.2
19. c   Section 41.2
20. b   Section 41.3

■ **Chapter 42 Test Answers**

1. c   Section 42.1
2. e   Section 42.1
3. a   Section 42.1
4. k   Section 42.1
5. j   Section 42.1
6. h   Section 42.2
7. f   Section 42.2
8. i   Section 42.2
9. d   Section 42.2
10. b   Section 42.2
11. b   Introduction
12. d   Introduction
13. a   Section 42.1
14. a   Section 42.1
15. c   Section 42.2
16. d   Section 42.2
17. c   Section 42.2
18. b   Section 42.2
19. a   Section 42.1
20. b   Section 42.2

■ **Chapter 43 Test Answers**

1. m   Section 43.1
2. e   Section 43.2
3. i   Section 43.2
4. h   Section 43.2
5. g   Section 43.2
6. c   Section 43.2
7. j   Section 43.2
8. f   Section 43.2
9. a   Section 43.2
10. d   Section 43.2
11. b   Section 43.2
12. k   Section 43.2
13. c   Section 43.1
14 b   Section 43.1
15. b   Section 43.1
16. c   Section 43.1
17. c   Section 43.2
18. a   Section 43.2
19. a   Section 43.2
20. d   Section 43.2

■ **Chapter 44 Test Answers**

1. h   Section 44.2
2. c   Section 44.1
3. l   Section 44.2
4. d   Section 44.1
5. j   Section 44.2
6. i   Section 44.2
7. a   Section 44.1
8. e   Section 44.2
9. k   Section 44.2
10. b   Section 44.2
11. m   Section 44.2
12. g   Section 44.2
13. b   Section 44.1
14. b   Section 44.1
15. c   Section 44.1
16. d   Section 44.2
17. d   Section 44.1
18. c   Section 44.1
19. a   Section 44.2
20. b   Section 44.2

■ **Chapter 45 Test Answers**

1. d   Section 45.1
2. i   Section 45.1
3. g   Section 45.1
4. k   Section 45.1
5. a   Section 45.1
6. h   Section 45.1
7. e   Section 45.2
8. j   Section 45.2
9. f   Section 45.2
10. b   Section 45.2
11. b   Section 45.1
12. a   Section 45.2
13. d   Section 45.2
14. c   Section 45.2
15. c   Section 45.1
16. a   Section 45.2
17. d   Section 45.2
18. b   Section 45.1
19. b   Section 45.2
20. d   Section 45.2

Answer Key for Carpentry Math

Math Activity Sheets for Chapter 1

M1-1
1. 292
2. 1,298
3. 16,599
4. 46
5. 434
6. 12,155
7. 744
8. 12,925
9. 64,158
10. 54
11. 92
12. 74
13. 1,213 sq. ft.
14. 50'
15. 965 sq. ft.
16. $528
17. $2,122
18. 14 gal.
19. $15,869
20. 1,170'

Math Activity Sheets for Chapter 2

M2-1
1. 9.6
2. 9.06
3. 9.006
4. 55.46
5. 68.024
6. 0.339
7. 0.0075
8. 726.058
9. 4.3208
10. 204.5
11. 67.5905
12. 194.8438
13. 2.3729
14. 38.7257
15. 4,372.5184
16. 15.2813
17. 16.1719
18. 0.4453
19. 25.0

M2-2
1. 902.912
2. 41.752
3. 563.25
4. 37.665
5. 161.76
6. 6.552
7. 0.406
8. 7.3
9. $49,132.19
10. $1,556.19
11. 101.5"
12. 13 treads
13. 89.4375"
14. $303.00
15. 7.375"
16. $176.94

Math Activity Sheets for Chapter 3

M3-1
A. $\frac{1}{16}$"
B. $\frac{1}{2}$"
C. $\frac{7}{8}$"
D. 1 $\frac{1}{4}$"
E. 1 $\frac{9}{16}$"
F. 2"
G. 2 $\frac{3}{8}$"
H. 2 $\frac{3}{4}$"
I. 3 $\frac{3}{16}$"
J. 3 $\frac{5}{8}$"
K. 3 $\frac{15}{16}$"
L. 4 $\frac{5}{16}$"
M. 4 $\frac{11}{16}$"
N. 5 $\frac{1}{8}$"
O. 5 $\frac{7}{16}$"
P. 5 $\frac{13}{16}$"
AA. 3'-9"
BB. 3'-9 $\frac{3}{8}$"
CC. 3'-9 $\frac{13}{16}$"
DD. 3'-10 $\frac{1}{4}$"
EE. 3'-10 $\frac{5}{8}$"
FF. 3'-11 $\frac{1}{16}$"
GG. 3'-11 $\frac{1}{2}$"
HH. 3'-11 $\frac{7}{8}$"
II. 4'-0 $\frac{5}{16}$"
JJ. 4'-0 $\frac{3}{4}$"
KK. 4'-1 $\frac{1}{8}$"
LL. 4'-1 $\frac{9}{16}$"
MM. 4'-2"
NN. 4'-2 $\frac{7}{16}$"
OO. 4'-2 $\frac{15}{16}$"
PP. 4'-3 $\frac{3}{16}$"
QQ. 4'-3 $\frac{11}{16}$"

M3-2
1. A. 0'-11 $\frac{1}{2}$"
   B. 0'-9"
   C. 0'-5 $\frac{1}{2}$"
   D. 0'-3"
   E. 0'-0 $\frac{1}{2}$"
   F. 1'-0"
   G. 2'-0"
   H. 3'-0"
   I. 5'-0"
   J. 7'-0"
   K. 8'-0"
   L. 9'-0"
2. A. 0'-10"
   B. 0'-6"
   C. 0'-2"
   D. 3'-0"
   E. 6'-0"
   F. 9'-0"
   G. 15'-0"
   H. 18'-0"
   I. 20'-0"
   J. 25'-0"
   K. 31'-0"
   L. 35'-0"

M3-3
1. A. 0'-10 $\frac{1}{4}$"
   B. 0'-4"
   C. 1'-0"
   D. 2'-0"
2. A. 0'-7"
   B. 3'-0"
   C. 8'-0"
   D. 11'-0"
3. A. 8'-11"
   B. 5'-6"
   C. 3'-4 $\frac{1}{2}$"
   D. 1'-1"
4. A. 14'-10"
   B. 11'-7"
   C. 9'-2"
   D. 4'-6"
5. A. 5'-10 $\frac{1}{2}$"
   B. 3'-8"
   C. 2'-3"
   D. 0'-6"
6. A. 11'-11"
   B. 8'-6"
   C. 5'-3"
   D. 2'-0"

Math Activity Sheets for Chapter 4

M4-1
1. $\frac{5}{16}$
2. $\frac{7}{8}$
3. $\frac{1}{4}$
4. $\frac{1}{2}$
5. $\frac{1}{4}$
6. $\frac{3}{4}$
7. $\frac{6}{8}$
8. $\frac{28}{32}$
9. $\frac{60}{64}$
10. $\frac{8}{16}$
11. $\frac{20}{64}$
12. $\frac{6}{16}$
13. 3 $\frac{3}{4}$
14. 1 $\frac{5}{8}$
15. 2 $\frac{5}{16}$
16. 4 $\frac{1}{2}$
17. 2 $\frac{11}{32}$
18. 2 $\frac{1}{64}$

M4-2
1. 35 $\frac{9}{16}$
2. 43 $\frac{9}{16}$
3. 248 $\frac{9}{32}$
4. 6 $\frac{5}{8}$
5. 33 $\frac{3}{16}$
6. 78 $\frac{3}{32}$
7. 142 $\frac{15}{16}$"
8. 29 $\frac{9}{16}$"
9. 18 $\frac{3}{4}$ hrs.
10. 30 $\frac{11}{16}$"

Math Activity Sheets for Chapter 5

M5-1
1. $\frac{27}{8}$
2. $\frac{21}{4}$
3. $\frac{15}{2}$
4. $\frac{99}{16}$
5. $\frac{35}{4}$
6. $\frac{41}{16}$
7. 4 $\frac{1}{2}$
8. 12 $\frac{3}{4}$
9. $\frac{7}{16}$
10. 6 $\frac{3}{8}$
11. 10 $\frac{1}{2}$
12. 15
13. 4 $\frac{1}{8}$
14. 73 $\frac{1}{2}$
15. 1 $\frac{1}{2}$
16. 2 $\frac{1}{16}$
17. 79 $\frac{3}{4}$
18. 21

M5-2
1. $\frac{2}{3}$
2. $\frac{3}{4}$
3. $\frac{1}{6}$
4. 3 $\frac{1}{2}$
5. $\frac{10}{11}$
6. 1 $\frac{3}{5}$
7. 118"
8. 180 $\frac{3}{4}$"
9. 13 treads
10. 94 $\frac{1}{8}$"
11. 7 $\frac{3}{16}$"
12. 40 courses
13. 61 $\frac{1}{8}$"

Carpentry & Building Construction Instructor Resource Guide
Copyright © Glencoe/McGraw-Hill

ANSWER KEY FOR CARPENTRY MATH (Continued)

Math Activity Sheet for Chapter 6

M6-1
1. 11'
2. 264"
3. 6'-6"
4. 39'
5. 9 yd. 1 ft.
6. 515"
7. 26'
8. 20 qt.
9. 15 gal. 2 qt.
10. 1,040 oz.
11. 42 weeks
12. 16,000 lb.
13. 13 lb. 9 oz.
14. 8 sq. ft.
15. 522 sq. ft.
16. 9 cu. ft.
17. 326,592 cu. in.
18. 896 sq. in.
19. 22 sq. ft. 58 sq. in.
20. 51 cu. ft. 1,415 cu. in.
21. 12,582 cu. in.

Math Activity Sheets for Chapter 7

M7-1
1. 46 yd. 1 ft. 4 in.
2. 14 cu. yd. 10 cu. ft. 700 cu. in.
3. 29 gal. 3 qt. 1 pt.
4. 81 wk. 5 days 3 hr.
5. 300 lb. 9 oz.
6. 247 sq. yd. 4 sq. ft. 2 sq. in.
7. 8 ft. 4 in.
8. 25 lb. 8 oz.
9. 2 yd. 2 ft.
10. 6 cu. yd. 8 cu. ft. 887 cu. in.
11. 2 gal. 2 qt.
12. 38 wk. 3 days 15 hr.
13. 3 tons 35 lb. 12 oz.
14. 269 sq. ft. 56 sq. in.
15. 1 qt. 1 pt.

M7-2
1. 35'-6"
2. 231 gal. 0 qt. 0 pt.
3. 31 days 2 hr. 50 min.
4. 63 cu. ft. 45 cu. in.
5. 199 yd. 1 ft. 0 in.
6. 153 lb. 2 oz.
7. 5'-2"
8. 4 gal. 3 qt.
9. 3 hr. 10 min. 46 sec.
10. 9 lb. 10 oz.
11. 3 yd. 2 ft. 5 in.
12. 6 cu. yd. 11 cu. ft.
13. 2 sq. yd. 5 sq. ft. 115 sq. in.
14. 9 wk. 5 days
15. 1 ton 1,400 lb.
16. 4'-8 ½"

Math Activity Sheet for Chapter 8

M8-1
1. 10'-0"
2. 8'-6"
3. 0'-3"
4. 0'-1 ½"
5. 3'-1"
6. 5'-4"
7. 2'-10"
8. 5'-7"
9. 3'-7 ½"
10. 6'-9 ½"
11. 11'-11"
12. 7'-11 ½"
13. 3'-2 ¼"
14. 5'-4 5/16"
15. 4'-2 ⅝"
16. 2'-4 15/16"
17. 1'-5 ¾"
18. 7'-1 11/16"
19. 3'-7 19/32"
20. 5'-9 ¾"
21. 7'-6 15/16"
22. 9'-10 9/16"
23. 2'-8 27/32"
24. 5'-2 11/16"

Math Activity Sheets for Chapter 9

M9-1
1. C
2. H
3. G
4. F
5. I
6. A
7. D
8. J
9. E
10. B
11. 35/100 = 7/20
12. 75/100 = ¾
13. 3/40
14. 3/400
15. 7 1/10
16. 5 1/1000
17. 58/1000 = 29/500
18. 8 125/1000 = 8 ⅛
19. 8 1/80
20. 3 8/100 = 3 2/25
21. 0.875
22. 0.046875 ~ 0.0469
23. 0.6875
24. 0.3
25. 0.023
26. 0.96875 ~ 0.9688

M9-2
1. 0.9688
2. 0.7344
3. 0.375
4. 0.1719
5. 0.75
6. 0.5625
7. 0.4688
8. 0.0625
9. 21/64
10. 5/64
11. ⅞
12. 11/16
13. 17/32
14. 3/16
15. 51/64
16. 15/64
17. 12 ½"; 12 ¼"; 12 ⅜"; 12 5/16"
18. 26"; 25 ¾"; 25 ¾"; 25 13/16"
19. 10"; 10"; 9 ⅞"; 9 15/16"
20. 13/16"
21. 21/32"
22. 96 9/32"
23. 6 25/32"

Math Activity Sheets for Chapter 10

M10-1
1. 68.75%
2. 66.67%
3. 108%
4. 27.3%
5. 0.53
6. 0.004
7. 0.2858
8. 0.9875
9. 117.19%
10. 10%
11. 0.0886
12. 0.0038
13. 265.42%
14. 79.65%
15. 1.5637
16. 0.0575
17. 3/32; **0.0938**; **9.38%**
18. ¾; **0.75**; 75%
19. 5/16; 0.3125; **31.25%**
20. 3/10; **0.3**; **30%**
21. 3 ⅘; **3.8**; 380%
22. 43/50; 0.86; **86%**
23. 5 ⅞; **5.875**; **587.5%**
24. ⅜; 0.375; **37.5%**
25. 1/50; **0.02**; 2%

M10-2
1. 10.25%
2. 7.6575
3. 274.33%
4. 12.5333
5. 76.83%
6. 55.7738
7. 136.88%
8. 129.1267
9. 0.125
10. 0.0838
11. 0.7188
12. 0.8594
13. 0.6667
14. 0.0013
15. 0.0025
16. 0.0009

	Fraction	Decimal	Percentage
17.	1/32	**0.0313**	**3.13%**
18.	½	**0.5**	50%
19.	⅚	0.8 ⅓	**83.33% or 83 ⅓%**
20.	⅖	**0.4**	**40%**
21.	2 ⅖	**2.4**	240%
22.	⅞	0.875	**87.5% or 87 ½%**
23.	5 ⅛	**5.125**	**512.5% or 512 ½%**
24.	3/20	0.15	**15%**
25.	1/25	**0.04**	4%

ANSWER KEY FOR CARPENTRY MATH (Continued)

Math Activity Sheets for Chapter 11

M11-1
1. 289 hours
2. 20%
3. $152,000
4. 28%
5. 1,989 members
6. $26.85

M11-2
1. $65.97
2. $142.96
3. $921.29
4.

Qty.	Item	Unit Price	Total Price
5	Framing Hammer	$19.97	$99.85
6	25' Tape Rule	$19.97	$119.82
2	Miter Boxes with Saw	$16.98	$33.96
4	48" Level	$23.88	$95.52
3	Stud Sensor	$11.13	$33.39
	SUBTOTAL		$382.54
	Less 20%, 8%		$281.55
	Sales Tax at 5 ¾%		$16.19
	TOTAL		$297.74

Math Activity Sheets for Chapter 12

M12-1
1. 354 in.
2. 8 ft.
3. 8 yd.
4. 56.5488 ft.
5. 78 ft.
6. 850 in.
7. 25.1328 yd.
8. 48 ft.

M12-2
1. 187.5 sq. ft.
2. 6.7083 sq. ft.
3. 36 sq. yd.
4. 50.2656 sq. ft.
5. 1,344 sq. in.
6. 5,832 sq. ft.

M12-3
1. 11.3906 cu. ft.
2. 5,038,848 cu. in.
3. 12.5664 cu. yd.
4. 336 cu. ft.
5. 19,200 cu. ft.
6. 637.5936 cu. in.

Math Activity Sheet for Chapter 13

M13-1
A. 1'-4 ½"
B. 1'-6"
C. 1'-7 ⅝"
D. 1'-9 ¼"
E. 1'-10 ⅛"
F. 2'-0"
G. 2'-1 ⅜"
H. 2'-2 ½"
I. 2'-3 ¾"
J. 2'-5 ⅞"
AA. 8.3'
BB. 8.47'
CC. 8.55'
DD. 8.62'
EE. 8.69'
FF. 8.76'
GG. 8.84'
HH. 8.91'
II. 9.03'
JJ. 9.08'

Math Activity Sheets for Chapter 14

M14-1
1. 4 yd.³
2. 1 ½ yd.³
3. 2 ½ yd.³
4. 6 yd.³
5. 7 ¼ yd.³
6. 15 ¾ yd.³
7. 1 yd.³
8. 13 ¾ yd.³
9. 8 ¼ yd.³
10. 468 ¾ yd.³

M14-2
1. A. 1,056
   B. 1,078
   C. 31
   D. 4
2. A. 420
   B. 429
   C. 12
   D. 1 ½
3. A. 234
   B. 239
   C. 7
   D. 1
4. A. 462
   B. 472
   C. 14
   D. 1 ¾
5. A. 1,311
   B. 1,338
   C. 38
   D. 4 ¾

Math Activity Sheets for Chapter 15

M15-1
1. 1 : 3 or ⅓
2. 1 : 50 or 1/50
3. 3 : 1 or 3/1
4. 3 : 7 or 3/7
5. 1 : 2 or ½
6. 4 : 1 or 4/1

M15-2
1. cement: 2; sand: 5, stone 8
2. cement: 3; sand: 5.25; stone 12
3. cement: 0.5; sand: 1; stone: 1.5
4. cement: 6 shovels; sand: 12 shovels; stone: 18 shovels
5. cement: 8 cu. ft.; sand: 18 cu. ft.; stone: 24 cu. ft.
6. 2 lb.
7. cement: 3 buckets; sand: 9 buckets; stone: 18 buckets

M15-3
1. cubic yards: 16 ¾; cost: $1,005.00; time: 7 ¼ hr.
2. cubic yards: 7 ½; cost: $472.50; time: 3 ¼ hr.
3. cubic yards: 3 ¾; cost: $278.13; time: 1 ¾ hr.
4. cubic yards: 8 ½; cost: $723.00; time: 3 ¾ hr.
5. cubic yards: 4; cost: $280.00; time: 1 ¾ hr.
6. cubic yards: 1 ¾; cost: $147.63; time: 1 hr.
7. cubic yards: 5; cost: $300.00; time: 2 ¼ hr.
8. cubic yards: 2; cost: $166.00; time: 1 hr.
9. cubic yards: 4 ¼; cost: $267.75; time: 2 hr.
10. cubic yards: 11 ½; cost: $747.50; time: 5 hr.

Math Activity Sheets for Chapter 16

M16-1
1. 1 ½"
2. ¾"
3. 3"
4. 1 ½"
5. 2 × 6
6. 9 ½"
7. 11"
8. 2"
9. 5 ½"
10. 20'
11. 15
12. 6"
13. 7"
14. 7 ½"
15. 3-2 × 4 × 14'
16. 28-2 × 8 × 16'
17. 3"
18. 14 ½"
19. 22 ½"
20. 10 ½"

M16-2
1. 128 bd. ft. 192 ft.
2. 560 bd. ft. 336 ft.
3. 118 bd. ft. 176 ft.
4. 272 bd. ft. 272 ft.
5. 720 ft.
6. 420 bd. ft.
7. 117 bd. ft.
8. 24 bd. ft.

ANSWER KEY FOR CARPENTRY MATH *(Continued)*

■ **Math Activity Sheet for Chapter 17**

M17-1
1. $219.52
2. $97.50
3. $54.25
4. $4,392.50
5. $419.84
6. cost: $456.00; board feet: 160
7. cost: $672.00; board feet: 480
8. cost: $308.75; board feet: 650

■ **Math Activity Sheets for Chapter 18**

M18-1
1. 154 sheets
2. 74 squares
3. 2,729 bricks
4. 138 sq. ft.
5. 185 ft.
6. 150 ft.
7. 280 ft.
8. 801 blocks
9. 59 sheets
10. 9 rolls

M18-2
1. actual sq. ft.: 168; waste added: 176.4 sq. ft.; no. of sheets: 6; cost: $111.36
2. actual sq. ft.: 256; waste added: 268.8 sq. ft.; no. of sheets: 9; cost: $191.16
3. actual sq. ft.: 1,198.3; waste added: 1,258.25 sq. ft.; no. of sheets: 40; cost: $547.60
4. actual sq. ft.: 1,568; waste added: 1,646.4 sq. ft.; no. of sheets: 52; cost: $606.32
5. actual sq. ft.: 1,472; waste added: 1,545.6 sq. ft.; no. of sheets: 49; cost: $732.55
6. actual sq. ft.: 1,872; waste added: 1,965.6 sq. ft.; no. of sheets: 62; cost: $1,212.72
7. actual sq. ft.: 2,136; waste added: 2,242.8 sq. ft.; no. of sheets: 71; cost: $609.18

■ **Math Activity Sheet for Chapter 19**

M19-1
1. 5'-3 ⅝"
2. 38'-2 3/16"
3. 2'-6"
4. 9'-2 5/16"
5. 1'-7 13/16"
6. 7'-6"

■ **Math Activity Sheets for Chapter 20**

M20-1

1. Floor Materials Estimate

Quantity	Unit	Stock Size	Description of Material	Unit Price	Total Price
7	posts	3 ½"	Steel lally columns	$59.48	**$416.36**
17	lengths	2 × 6 × 14'	Sill stock	$5.92	**$100.64**
5	rolls	¼" × 5 ½" × 50'	Sill sealer	$4.19	**$20.95**

2. Floor Materials Estimate

Quantity	Unit	Stock Size	Description of Material	Unit Price	Total Price
12	posts	3 ½"	Steel lally columns	$59.48	**$713.76**
18	lengths	2 × 4 × 16'	Sill stock	$5.01	**$90.18**
6	rolls	¼" × 3 ½" × 50'	Sill sealer	$2.79	**$16.74**

M20-2

1. Floor Materials Estimate

Quantity	Unit	Stock Size	Description of Material	Unit Price	Total Price
73	lengths	2 × 10 × 20'	Joist stock	$18.77	**$1,370.21**

2. Floor Materials Estimate

Quantity	Unit	Stock Size	Description of Material	Unit Price	Total Price
169	lengths	2 × 8 × 20'	Joist stock	$15.87	**$2,682.03**

M20-3

1. Floor Materials Estimate

Qty	Unit	Stock Size	Description of Material	Unit Price	Total Price
72	lengths	1 × 3 × 8'	Cross Bridging Stock	$1.78	**$128.16**
80	sheets	⅝" × 4' × 8'	Plywood Sheathing	$14.95	**$1,196.00**

2. Floor Materials Estimate

Qty	Unit	Stock Size	Description of Material	Unit Price	Total Price
120	lengths	1 × 3 × 8'	Cross Bridging Stock	$1.78	**$213.60**
134	sheets	½" × 4' × 8'	Plywood Sheathing	$11.66	**$1,562.44**

ANSWER KEY FOR CARPENTRY MATH (Continued)

Math Activity Sheets for Chapter 21

M21-1
1. A. 256 studs
   B. $568.32
   C. 111/8s
   D. $251.97
   E. 12/14s
   F. $202.08
   G. $1,022.37
2. A. 120 studs
   B. $420.00
   C. 35/16s
   D. $277.20
   E. 10/14s
   F. $90.50
   G. $787.70
3. A. 216 studs
   B. $1,021.68
   C. 63/12s
   D. $340.20
   E. 3/10s
   F. $32.10
   G. $1,393.98
4. A. 184 studs
   B. $754.40
   C. 46/14s
   D. $272.32
   E. 15/12s
   F. $131.40
   G. $1,158.12

M21-2
1. 43'-3 3/16"
2. 31'-7 1/2"

Math Activity Sheets for Chapter 22

M22-1
1. 1/2
2. 1/4
3. 3/4
4. 1/6

M22-2
1. 12/12
2. 3/12
3. 16/12
4. 3/12

M22-3

Pitch	Slope
1/2	2/12
1/8	3/12
1/6	4/12
5/24	5/12
1/4	6/12
7/24	7/12
1/3	8/12
3/8	9/12
5/12	10/12
1/2	12/12
5/8	15/12
3/4	18/12

1. pitch: 1/4; slope: 6/12
2. pitch: 1/2; slope: 12/12
3. pitch: 3/4; slope: 18/12
4. rise: 4'; slope: 3/12
5. rise: 2'-8"; pitch: 1/6

Math Activity Sheet for Chapter 23

M23-1
1. line length: 23'-10"; less the ridge board: 23'-9 1/4"
2. line length: 32'-3 1/4"; less the ridge board: 32'-2 1/2"
3. hip length: 15'-3 5/8"
4. rafter length: 16'-4 1/16"; less the ridge board: 16'-3 5/16"

Math Activity Sheet for Chapter 24

M24-1
1. 21'-7 5/8"
2. 16'-9 3/16"
3. 24'-2 13/16"
4. 21'-4 5/16"
5. 20'-1 15/16"
6. 13'-6 15/16"
7. 14'-4 7/8"
8. 23'-0"
9. 12'-9 5/8"
10. 43'-3 3/16"

Math Activity Sheet for Chapter 25

M25-1
1. 90 rafters
2. 46 rafters
3. 29 trusses
4. 37 trusses
5. 44-2 × 8 × 22' and 44-2 × 8 × 18'
6. 8 rafters
7. 60-2 × 8 × 20' and 26-2 × 8 × 16'
8. 425 trusses
9. 1,512 rafters
10. 52-2 × 10 × 22' and 52-2 × 10 × 18' rafters

Math Activity Sheets for Chapter 26

M26-1
1. 42'-8 15/16"; 42.7448'
2. 16'-5 11/16"; 197 11/16"
3. 12'-6"
4. 22'-7 15/16"
5. 881.0749 sq. ft.; 126,874.78 sq. in.
6. 3,527.569 sq. ft.; 391.9521 sq. yd.
7. 715.5919 cu. ft.; 1,236,543 cu. in.
8. 65.2933 cu. ft.; 2.4183 cu. yd.
9. 0'-7 9/16"
10. 12

M26-2
1. 864 bd. ft.
2. $332.64
3. 4,620 bd. ft.
4. $2,217.60
5. $1,021.52

M26-3
1. 72.6024 sq. ft.
2. 79'-10 9/16"
3. 122.7185 sq. ft.
4. 58'-11 13/16"
5. area: 171.8399 sq. ft.; circumference: 46'-5 5/8"

M26-4
1. 9/12
2. 10'-0"
3. 15'-2"
4. 27'-4 1/8"
5. 27'-3 3/8"
6. 5'-5 1/4"
7. 18'-9 5/16"
8. 18'-8 9/16"
9. 28'-2 13/16"
10. 3/12
11. 16'-5 15/16"
12. 18'-6 5/8"
13. 10'-8 13/16"
14. 24'-0 1/16"
15. 25'-4 13/16"

Math Activity Sheet for Chapter 27

M27-1
1. 24" OC
2. 16" OC
3. A. 23'-0 3/8"
   B. 21'-0 3/8"
   C. 19'-0 3/8"
   D. 17'-0 3/8"
   E. 15'-0 3/8"
   F. 13'-0 3/8"
   G. 11'-0 3/8"
   H. 9'-0 3/8"
   I. 7'-0 3/8"
   J. 5'-0 3/8"
   K. 3'-0 3/8"
   L. 1'-0 3/8"
4. 24
5. 18/12
6. 30'-1"
7. 18'-3 11/16"
8. 10'-11 13/16"
9. 23'-5 5/16"
10. A. 16'-7 11/16"
    B. 14'-11 11/16"
    C. 13'-3 11/16"
    D. 11'-7 11/16"
    E. 9'-11 11/16"
    F. 8'-3 11/16"
    G. 6'-7 11/16"
    H. 4'-11 11/16"
    I. 3'-3 11/16"
    J. 1'-7 11/16"
11. 20

ANSWER KEY FOR CARPENTRY MATH *(Continued)*

Math Activity Sheet for Chapter 28

M28-1

Material Estimate

Qty	Unit	Size	Description of Material	Unit Price	Total Price
14	windows	32 × 58	**Double hung DH3258**	$226.00	**$3,164.00**
3	windows	46 × 36	**Sliders SL4636**	$118.00	**$354.00**
3	doors	32 × 80	**Hollow core HC3280**	$32.70	**$98.10**
5	doors	30 × 80	**Hollow core HC3080**	$29.70	**$148.50**
				Subtotal	**$3,764.60**
				Less 15% Discount	**$564.69**
				Subtotal	**$3,199.91**
				5 ¾% Tax	**$183.99**
				Total	**$3,383.90**

Math Activity Sheets for Chapter 29

M29-1
1. T
2. F
3. T
4. T
5. $x = 26.6667$
6. $a = 2.7368$
7. $c = 10$
8. $y = 0.3175$
9. $d = 3.0588$
10. $x = 285$
11. $z = 1$
12. $m = 0.1944$
13. $n = 1,118.25$
14. $p = 10$
15. $s = 180.5263$
16. $t = 10.9091$
17. $x = 950.6977$
18. $n = 14,012$
19. $y = 20$
20. $z = 12.5667$

M29-2
1. $d = \$10.67$
2. $f = 6.4$ sq. ft.
3. $d = 18.6667$ days
4. $x = \$128.00$
5. 87.5 hrs.
6. 15 hrs.
7. 22.5 hrs.
8. $133.25
9. 6 lbs.
10. 7.2 ~ 8 gal.
11. 86.4 ~ 87 bundles
12. 1,687.5 ~ 1,688 brick
13. 25.1429 ~ 26 rolls
14. 3.8804 ~ 4 rolls
15. $41.93

M29-3
1. 6 days
2. 13.3333 days
3. 12 days
4. 65 rpm
5. 3 days
6. 110 min. or 1 hr. 50 min.
7. 9 hrs.
8. 7 ½ hrs.
9. 96 rpm
10. 4"

M29-4
1. 124.416 hrs.
2. 378.624 hrs.
3. 62.08 hrs.
4. 85.12 hrs.
5. 270.6 hrs.
6. 920.84 hrs.
7. 231 hrs.

Labor Calculation Table

Worker	Hours	Rate/Hour	Gross Pay
Supervisor	**231**	$19.50	**$4,504.50**
Carpenter	**231**	$12.50	**$2,887.50**
Helper	**231**	$8.00	**$1,848.00**
Laborer	**231**	$6.50	**$1,501.50**

ANSWER KEY FOR CARPENTRY MATH *(Continued)*

Math Activity Sheet for Chapter 30

M30-1

1.

Material Estimate

Qty	Unit	Stock Size	Description of Material	Unit Price	Total Price
2	roll	10" × 10'	aluminum flashing	$5.69	**$11.38**
23	length	3" × 4" × 10'	galvanized drip edge	$3.25	**$74.75**
8	length	13 ¾" × 8'	aluminum/ridge vent	$8.74	**$69.92**
7	roll	432 sq. ft.	15 lb. roofing felt	$11.98	**$83.86**
83	bundle	33 ⅓ sq. ft.	25 year arch. shingles	$11.98	**$994.34**
				Subtotal	**$1,234.25**
			Less 12% & 5% Discount		**$202.42**
				Subtotal	**$1,031.83**
				5 ¾% Tax	**$59.33**
				Total	**$1,091.16**

2.

Material Estimate

Qty	Unit	Stock Size	Description of Material	Unit Price	Total Price
1	roll	24" × 50'	aluminum flashing	$49.99	**$49.99**
16	length	4" × 4" × 10'	galvanized drip edge	$4.95	**$79.20**
1	roll	⅝" × 11 ¼" × 20'	flexible ridge vent	$37.99	**$37.99**
10	roll	216 sq. ft.	30 lb. roofing felt	$11.98	**$119.80**
63	bundle	33 ⅓ sq. ft.	40 year arch. shingles	$15.25	**$960.75**
				Subtotal	**$1,247.73**
			Less 10% & 2% Discount		**$147.23**
				Subtotal	**$1,100.50**
				5 ¾% Tax	**$63.28**
				Total	**$1,163.78**

Math Activity Sheets for Chapter 31

M31-1
1. number of lengths: 36; cost: $107.64

M31-2
1. number of lengths: 16; cost: $47.84

M31-3
1. 6 outside
   2 inside
   $65.92
2. 5 outside
   1 inside
   $50.94

M31-4
1. number of lengths: 22; cost: $118.58

Math Activity Sheets for Chapter 32

M32-1
1. number of lengths: 23; cost: $68.77
2. number of lengths: 24; cost: $71.76

M32-2
1. number of squares: 21; cost: $923.79; pounds of nails: 21
2. number of squares: 20; cost: $879.80; pounds of nails: 20

M32-3
1. number of lengths: 34; cost: $288.66
2. number of lengths: 34; cost: $288.66

M32-4
1. number of lengths: 22; cost: $147.18
2. number of lengths: 23; cost: $153.87

Math Activity Sheets for Chapter 33

M33-1
1. cement: 7 cu. ft.; lime: 3 ½ cu. ft.; sand: 31 ½ cu. ft.
2. cement: 5 cu. ft.; lime: 1 ¼ cu. ft.; sand: 15 cu. ft.
3. cement: 2 cu. ft.; lime: 1 cu. ft.; sand: 9 cu. ft.
4. cement: 7 cu. ft.; lime: 7 cu. ft.; sand: 42 cu. ft.
5. cement: 13 cu. ft.; lime: 3 ¼ cu. ft.; sand: 39 cu. ft.

M33-2
1. actual brick: 2,848; ordered brick: 2,905; cement: 23 bags; sand: 3 tons; hours: 36
2. actual brick: 4,212; ordered brick: 4,297; cement: 34 bags; sand: 4 ½ tons; hours: 53
3. actual brick: 2,366; ordered brick: 2,414; cement: 19 bags; sand: 2 ½ tons; hours: 30
4. actual brick: 5,184; ordered brick: 5,288; cement: 42 bags; sand: 5 ¼ tons; hours: 65
5. actual brick: 4,954.5; ordered brick: 5,054; cement: 40 bags; sand: 5 tons; hours: 62

ANSWER KEY FOR CARPENTRY MATH (Continued)

Math Activity Sheets for Chapter 34

M34-1
1. total rise: 98.25"; no. of risers: 14; unit rise: 7"; no. of treads: 13; unit run: 10 ½"; total run: 11'-4 ½"
2. total rise: 114.375"; no. of risers: 16; unit rise: 7 ⅛"; no. of treads: 15; unit run: 10 ⅜"; total run: 12'-11 ⅝"

M34-2
1. stringer length: 14'-0 ⅛"; length to order: 16s; drop-off amount: ⅞"
2. stringer length: 16'-1 ⅛"; length to order: 18s; drop-off amount: ⅝"
3. stringer length: 8'-5 13/16"; length to order: 10s; drop-off amount: 9/16"
4. stringer length: 5'-9 1/16"; length to order: 8s; drop-off amount: ¼"
5. stringer length: 13'-10 15/16"; length to order: 16s; drop-off amount: ⅜"
6. stringer length: 3'-6 ¼"; length to order: 8s; drop-off amount: 5/16"

M34-3
1. $335.75
2. $414.38
3. $430.50
4. $507.50

M34-4
1. 14 risers
2. 7 ½"
3. −0 3/16"
4. 13 treads
5. 10 15/16"
6. 0"
7. 13'-9 ⅞"
8. 36.08° ~ 36°

Math Activity Sheet for Chapter 35

M35-1
1. 86'
2. 54'
3. baseboard: 24'; shoe base: 24'
4. 22'
5. 184'
6. 248'

Math Activity Sheet for Chapter 36

M36-1
1. 20 sq. ft.; $31.00
2. 15 sq. ft.; $21.75
3. 45 sq. ft.; $78.75

Math Activity Sheet for Chapter 37

M37-1
1. 1.087
2. 1.111
3. 1.136
4. 1.163
5. 1.190
6. $12,260.35
7. $32,343.75
8. $33,938.31

Math Activity Sheets for Chapter 38

M38-1
1. 31 ⅛"
2. 24 1/16"
3. 22 ¼"
4. piping: 34"; longer leg: 29 7/16"
5. piping: 39 ½"; longer leg: 34 3/16"
6. piping: 26 ¼"; longer leg: 22 ¾"

M38-2
1. 20 amps
2. 14 ohms
3. 7 amps
4. 202.125 volts
5. 16 ohms
6. 105 volts

M38-3
1. Exterior walls: 9,120 Btu; Windows/doors: 9,600 Btu; Ceiling: 10,658 Btu; Floor: 0 Btu; Total: 29,378 Btu
2. Exterior walls: 28,917 Btu; Windows/doors: 16,866.679 Btu; Ceiling: 13,908.125 Btu; Floor: 0 Btu; Total: 59,692 Btu

Math Activity Sheet for Chapter 39

M39-1
1. 32 bags; $223.36
2. 37 rolls; $3,552.00
3. 906 rolls; $16,561.68
4. 1,199 rolls; $21,917.32
5. 50 bags; $349.00
6. 43 bags; $300.14
7. 775 sheets; $5,804.75

Math Activity Sheet for Chapter 40

M40-1
1. A. 65 sheets $258.70
   B. 4 rolls of tape $5.00
   C. 5 drums of compound $41.85
   D. Total Cost $305.55
2. A. 49 sheets $293.02
   B. 4 rolls of tape $5.00
   C. 6 drums of compound $50.22
   D. Total Cost $348.24
3. A. 55 sheets $218.90
   B. 3 rolls of tape $3.75
   C. 4 drums of compound $33.48
   D. Total Cost $256.13
4. A. 282 sheets $1,686.36
   B. 20 rolls of tape $25.00
   C. 31 drums of compound $259.47
   D. Total Cost $1,970.83

Math Activity Sheet for Chapter 41

M41-1
1. gallons: 13; $220.61
2. gallons: 2; $45.94
3. gallons: 3; $50.91
4. gallons: 4; $63.88
5. gallons: 5; $44.90
6. gallons of primer: 3; $26.97
   gallons of semi-gloss: 2; $37.94
7. gallons: 14; $279.58
8. 5-gallon drums: 9; $350.73

Math Activity Sheet for Chapter 42

M42-1
1. strips: 563; cases: 28; cost: $751.80
2. planks: 253; cases: 28; cost: $1,717.80
3. square yards: 47; cost: $951.75
4. A. tile: 52; cases: 7; cost: $107.52
   B. tile: 104; cases: 9; cost: $414.72
   C. planks: 154; cases: 7; cost: $720.72
5. A. tile: 124; cases: 4; cost: $313.92
   B. tile: 278; cases: 6; cost: $426.24
   C. sq. yd.: 31; cost: $835.45
   D. pavers: 1,029; cost: $648.27

ANSWER KEY FOR CARPENTRY MATH *(Continued)*

■ Math Activity Sheet for Chapter 43

M43-1
1. slates: 202; cost: $472.68
2. tiles: 454; cost: $304.18
3. pavers: 908; cost: $444.92
4. tiles: 172; cost: $602.00
5. tiles: 330; cost: $1,640.10
6. tiles: 147; cost: $320.46
7. tiles: 598; cost: $77.74
8. tiles: 308; cost: $40.04
9. tiles: 569; cost: $73.97
10. tiles: 254; cost: $33.02

■ Math Activity Sheet for Chapter 44

M44-1
1. 19 courses
2. 21 courses
3. 14 courses
4. 6 courses
5. 16 courses
6. 11 courses
7. 45 $5/16$"
8. 66"
9. 35 $3/16$"
10. 116"
11. 48"
12. 104"

■ Math Activity Sheet for Chapter 45

M45-1
1. posts: 8; on-center spacing: 4'-8"; cubic yards: 1
2. posts: 13; on-center spacing: 4'-4 $13/16$"; cubic yards: 2